Buddhism and Iconoclasm in East Asia

Also available from Bloomsbury

Chinese Religion, Yanxia Zhao and Xinzhong Yao

Buddhism and Iconoclasm in East Asia

A History

Fabio Rambelli and Eric Reinders

BLOOMSBURY
LONDON • NEW DELHI • NEW YORK • SYDNEY

Bloomsbury Academic
An imprint of Bloomsbury Publishing Plc

50 Bedford Square
London
WC1B 3DP
UK

175 Fifth Avenue
New York
NY 10010
USA

www.bloomsbury.com

First published 2012

British Library Cataloguing-in-Publication Data
A catalogue record for this book is available from the British Library.

ISBN: HB: 978-1-4411-4509-3
ePDF: 978-1-4411-8168-8

Library of Congress Cataloging-in-Publication Data
Rambelli, Fabio.
Buddhism and iconoclasm in East Asia : a history / Fabio Rambelli and Eric Reinders. – First [edition].
pages cm
Includes bibliographical references and index.
ISBN 978-1-4411-4509-3 (hardcover) – ISBN 978-1-4411-8168-8 (ebook (pdf)) 1. Buddhism and culture–East Asia. 2. Matter–Religious aspects–Buddhism. 3. Violence–Religious aspects–Buddhism. I. Reinders, Eric Robert. II. Title.
BQ616.R37 2012
294.3'437—dc23
2012010760

Typeset by Newgen Imaging Systems Pvt Ltd, Chennai, India
Printed and bound in Great Britain

The earlier culture will become a heap of rubble and finally
a heap of ashes, but spirits will hover over the ashes.

(Ludwig Wittgenstein)

Contents

Preface

This book began in a rather casual and unpredictable way some 15 years ago. One evening at dinner, after a long day spent listening to paper presentations at an annual meeting of the American Academy of Religion, we were musing on the fact that many paper titles included words that implied a constructivistic approach, such as "constructing," "creating," and "producing." Even "de-constructing" (an expression much present in conference papers back then) is predicated on a previous "construction"; other common terms, such as "engendering" and "embodying," also presuppose more or less intentional acts of making. Suddenly, all this struck us as slightly incongruous, because we were well aware that a large portion of religious practice deals not with construction but rather with *destruction*. It occurred to us that destruction within the religious field is mostly the topic of news (as in acts of religious fundamentalism or terrorism), and is usually reduced to actions of fanatics. There exists a more specialized academic literature on iconoclasm, but almost always limited to particular Western cases. By inverting many of the conference paper titles—from creating to destroying, doing to undoing, embodying to disembodying—we imagined a whole catalogue of destruction. Destruction now showed us two faces: one aimed at attacking religious institutions and representations, the other aimed at creating and preserving them.

We gradually developed a joint manuscript. We also decided to add "solo" chapters, written by each of us according to our respective areas of expertise and topics of interest. It became clear that our project could never be in any sense exhaustive, as the field of religious destruction in East Asia is vast and still largely unexplored. Rather, since the beginning we set for ourselves the following aims.

First, we wanted to draw attention to phenomena of religious destruction in East Asia that can be characterized as "iconoclastic" despite the received, narrowly Eurocentric understanding of the term. In order to do so, we had to expand the meaning of "iconoclasm" to include, literally, all forms of destruction of religious objects; this was our second goal. Third, we wanted to show that iconoclasm targets not just objects but also people—often according to the same logic and by following the same rules. Fourth, we proposed (rather ambitiously), a general catalogue of religious destruction, which, we hoped, may be useful also to study cases outside of the East Asian context. Finally, we envision the birth of a new discipline, "Destruction Studies," that would address any and all instances of destruction within cultures.

A comment about our terminology is in order here. Some readers will be puzzled, if not even annoyed, by our attempt to extend the term "iconoclasm" to East Asia; after all, Paul Demiéville seems to have closed the issue several decades ago when he wrote that destruction of sacred images in China was caused mostly by economic reasons and not by intellectual concerns (Demiéville, 1974). However, recent studies have emphasized the existence of a larger, pervasive dimension of Western iconoclastic movements, as not simply the result of doctrinal diatribes, but as complex phenomena related to economic, political, and social issues as well; furthermore, all these issues also involved discourses about the representation of the sacred. It is according to this new understanding that we apply the term iconoclasm to East Asian religious history.

The word "iconoclasm" is rich with meaning and history, and it has come to mean many different things, in different time periods and in different disciplines. We need to broaden the use of the term beyond certain historical associations and disciplinary conventions, in an attempt to define a trans-disciplinary object of analysis. We employ the term also in consideration that, so far, there is no better term. After all, if we bracket for a moment the thick Byzantine and Protestant connotations of the word, "iconoclasm" simply means "destruction of sacred images" (and related sacred objects). We do introduce, however, the term "hieroclasm" toward the end of the book, to point to a well-delimited phenomenon, namely, the elimination of sacred value from formerly religious artifacts once they are re-signified as cultural relics or art.

In our usage, iconoclasm is a subset of the larger category *destruction*. Though some acts of destruction (of things and people) are so sad or horrific as to demand not only the strongest criticism but also concerted efforts (including violent intervention) to prevent them, yet we do not in principle consider the category "destruction" as negative, morbid, or evil. The destruction of objects produces new meanings and practices, damaged things may become more precious, and the dead continue to influence us. In many ways, destruction is not the end of culture but one of the conditions of its possibility. Destruction *is* cultural activity. Let us not forget that production (of a wooden icon, say) always involves some degree of destruction (of a tree).

The destruction of religious objects is a cultural practice that changes the materiality or the meaning of the object involved, or both. Destruction and damage of religious objects cause transformations of the semiotic status of those objects. Operating on the materiality—on the body—of a sacred object affects and modifies its symbolic status—its meanings and functions in its cultural contexts. In certain cases, communities are willing to dispose of objects they consider no longer "sacred" or, at least, indispensable, to replace them either with new objects of identical use or with totally different objects carrying identical "value" (see Barbier, 1994). Destruction or damage of objects may also transform the status of the agents involved. In this respect, a promising new avenue of research has been opened by a number of scholars who have addressed acts of destruction (especially, the destruction of sacred objects) as ritual forms and, more generally, as signifying practices. Natalie Zemon Davis, for example, in her fascinating study of religious riots in premodern France, treated acts of destruction of religious objects as "texts" with an internal logic that can "be 'read' as fruitfully as a diary, a political tract, a sermon, or a body of laws" (Davis, 1975, pp. xvi–xvii). She showed that the violence of crowds against religious targets is not simply something so profoundly irrational as to be beyond the scope of research and interpretation. On the contrary, "violence, however cruel, [was] not . . . random and limitless but [was] aimed [instead] at defined targets and selected from a repertory of traditional punishments and forms of destruction" (ibid., p. 154).

The destruction of sacred objects in China and Japan gives a different picture of East Asian history and Buddhism. For instance, thinking about religious phenomena such as Christian missions as (among other things) traditions of destruction changes the historical narratives. In the case of Japan, to see the ways in which religious institutions presented visions of incessant cosmic struggle underlying concrete historical events offers a rather different picture of Japanese Buddhism. We are not aiming for an exhaustive account, something that would be impossible anyways. Our aim is both more modest and more ambitious than that. It is more modest in the sense that we simply hope to start a discussion on the relations between Buddhist, its materiality, and instances of religious violence and destruction. It is also more ambitious, however, because we are gesturing toward some kind of generalized theory of meaningful destruction. In the book, we focus on particular instances because the specificity and concreteness of acts of iconoclasm require a case-by-case approach. However, we shall also step back from the particulars and seek a cross-fertilization of theoretical approaches to destruction in culture. This book seeks to chart out variations in the field of iconoclasm in East Asia. It also deliberately addresses the boundaries or gray areas around the central concepts of iconoclasm. An interdisciplinary and cross-cultural examination of these and other issues will enable us to formulate some hypotheses on the mechanisms of symbolic destruction in religious and cultural phenomena. As we address a larger audience than Buddhist studies and East Asian religions specialists, we have tried to reduce the use of primary sources to the minimum indispensable, and we quote from translations whenever possible.

Let us now look at the structure of the book. It is divided into three parts, in an ideal itinerary from the more concrete (material objects) to the more abstract (theories about cultural value and destruction), from Buddhology (the material dimension of Buddhism), to East Asian historical and cultural contexts (Japan, China, and museums) and to cultural analysis in general. The first part, "Stuff: Materiality and Fragility of Dharma," lays out the main conceptual framework of the book. Chapter 1, "Buddhist Objects, Buddhist Bodies—An Outline," presents an overview of the field of Buddhist materiality and the keen awareness, within it, of its fragility and the ever-present possibility of destruction. We draw attention to the enormous importance of material objects for not just Buddhist practice, but also for the conceptualization of Buddhist doctrines; this importance is well indicated by the sheer number of objects directly involved in the discourses and practices of Buddhists throughout the ages. The chapter explores the entangled connections between the need for objects (in order to think and act in Buddhist terms) and the uneasiness toward such a need. We also discuss the pervasive rhetoric of animation, according to which Buddhist sacred images are real presences of Buddhas; this rhetoric contributed to blurring the lines separating objects and human bodies, something that affected also the ways in which iconoclastic attacks were carried out. However, Buddhists were also very aware that objects are perishable: subject to wear-and-tear, unintentional damage, and intentional acts of destruction. Indeed, Buddhists also believed that Buddhism itself was going to disappear one day, at the end of the Final Age of Dharma (Ch. *mofa*, Jp. *mappō*), and that the end would be accelerated by excessive attention to materiality. In the final section of the chapter we discuss the concept of iconoclasm and scholarly developments in the field that address not only theological issues but also social background, historical context, politics, and economics.

We look at these new developments, with their wider perspectives, to ground our own perspective on instances of destruction of sacred objects in East Asia. In particular, we criticize the received idea that there was not iconoclasm in East Asia because the intellectual traditions there did not formulate specific religious injunctions against image worship; we argue instead that destruction of sacred images is a meaningful act per se that questions anyway the status of the objects (especially, their sacredness) it targets.

The second part, "Histories: Instances of Religious Destruction in East Asia," consists of three chapters, each envisioned as a large case study of instances of iconoclasm broadly understood. Each chapter deals with a specific topic, associated either to geographical areas (Japan, China) or specific institutions and practices related to display and observation of sacred objects (museums, exhibitions, temple displays, and tourism). Chapter 2, "Iconoclasm and Religious Violence in Japan: Practices and Rationalizations," by Fabio Rambelli, begins with a discussion of some of the ways in which religious institutions in contemporary Japan present cases of destruction that have affected them in the past; next, it proposes some general historical trends in iconoclastic practices in Japan, distinguishing between medieval, early modern, and modern iconoclasm. What distinguishes them are the goals of destruction, the scale of destruction, and the situations in which intentional destruction was authorized. The chapter also addresses premodern documents in which history is described as a continuous state of struggle between Buddhism and its opponents, epitomized by a cosmic struggle opposing Māra to the Buddha. Within such an intellectual context, acts of destruction affecting Buddhist people, institutions, and objects acquire a completely new light. Chapter 3, "Shattered on the Rock of Ages: Western Iconoclasm and Chinese Modernity," by Eric Reinders, deals with Christian and Christian-inspired iconoclasm in China. The choice of this topic in the context of this book deserves an explanation. For the most part, this is not an "East-West" comparative book, nor do we hold the Christian iconoclasms to be the normative cases. However, it is important to recognize that, from the sixteenth century onwards, Christian iconoclastic acts and discourses began to be present in East Asia, and this chapter reaches back to those Western iconoclastic traditions to see how they interacted with Chinese culture. This chapter includes discussion of Jesuit objections to Buddhist icons, Protestant anti-Catholicism as a template to reject Chinese "idolatry," relations between iconoclasm and colonialism, the ceremonies of rejecting and destroying icons, and anti-Christian iconoclasm. The chapter further explores how iconoclasm became a major motif in Chinese modernism, such as in the way that icons became "trophies" of superstition and in the practice of using temple buildings as secular schools. In scientific, secular, and, even, Marxist hostility to popular cults one can see the continued influence of Christian missions. The chapter next outlines the patterns of Communist iconoclasm, focusing in more detail on the Cultural Revolution. During this period, some icons, scriptures, and temples were attacked, but by the mid-1960s not very much was left to destroy; accordingly, the final part of the chapter deals with attacks on human "icons" (people representing or forced to represent demonized values), in a discussion that returns to the theme of iconoclasm and the body. Since missionaries, the May Fourth movement, and Red Guards made little effort to distinguish between Buddhists and any other forms of "superstition," this chapter makes the least reference to Buddhism in particular. Next, in Chapter 4, "Ways of Not Seeing: Cultural Redefinition and Iconoclasm," written again by the two of us together, we explore what happens when objects are put on display in museums, and in which ways museum display is different from (or, at times, similar to) temple display, by focusing on various semiotic strategies employed (such as light, visibility, trajectories, and labeling). We understand museum display as a form of radical re-signification of the sacred. This raises the broader issue of re-signification for objects and bodies targeted by iconoclastic acts.

Part Three, "Theories: Rethinking the Relations Between the Sacred and Destruction," is the more theoretical portion of the book; it steps back from direct geographical and historical contexts and attempts to tackle with more general issues related to cultural value, sacredness, and destruction. Chapter 5, "Orders of Destruction: Iconoclasm, Semioclasm, Hieroclasm," offers a general catalogue of iconoclastic acts based on the harm done to, respectively, affected objects and human bodies. We show that destructive acts can be classified in terms of intentionality (or lack thereof), the ways they affect the objects involved (ranging from complete destruction to complete preservation but in a new context that changes the previous significance of the object), but also in positive or negative terms (received forms of iconoclasm negate the objects they target, but other forms in which objects are harmed or destroyed in a religious context, such as sacrifice or humility, aim at enhancing the objects they affect). Furthermore, we find a significant parallelism between ways in which iconoclastic acts target objects and human bodies; sometimes, objects are treated as animated bodies, sometimes bodies are de-humanized and treated as inanimate objects. We also propose a more general theoretical framework to understand instances in which the status and the meaning of the sacred in any given culture is questioned, contested, and ultimately denied. We present a typology of objects, discuss ways to shift the significance of an object from one typology to another, and address the semiotic labor involved in re-signification by showing how iconoclastic acts target not simply objects and related practices, but also and especially these objects' signifiers, signifieds, and the codes connecting them. In the Conclusion, "Destruction and Cultural Systems," we emphasize the multifarious roles of destruction in cultural activity and argue for the need of a new field of "destruction studies" that would address all the above issues.

It is important to make clear that we do not particularly *like* destruction, and by character and ideology, we are deeply nonviolent. Our interest in violence and destruction is not motivated by a desire to condone them. However, we feel that the popular explanations of destructive acts (as random acts of deranged individuals or angry mobs, expressions of religious fanaticism, or terrorist strategies) are woefully inadequate. We hope we can show that destruction is in fact a complex act, involving different intentionalities, agents, means, goals, and possible interpretations; also, most effective acts of destruction target not just the mere materiality (physical body) of sacred objects, but also and especially their semiotic structure (their meanings and the ways those meanings are embodied and experienced).

The third millennium opened with two shocking acts of iconoclasm (some aspects of their far-ranging impact have been discussed in Belpoliti 2002). In March 2001, the Talibans in Afghanistan destroyed the Great Buddhas at Bamiyan in Afghanistan, after having looted and destroyed much of the artistic and cultural patrimony of the country, including countless artifacts and artworks dating back to the early state of Buddhist iconic art (see Flood, 2002). Next, on September 11, terrorists attacked symbolically charged sites in the United States, in particular the Twin Towers of the World Trade Center. The second iconoclastic act, in particular, has had enormous consequences, to the point of shaping international relations and the way we perceive our world.

More generally, we are aware that destruction too can become a fetishized object of desire, whether in the suicide bomber's milieu, or in heavy metal music and Goth aesthetics. Just as the Buddhist analysis of things as innately susceptible to disintegration is not a "negative" or "unwholesome" starting point (as Pope John Paul II claimed), so too we did not write this book in any morbid or apocalyptic mood. The supposed "negativity" of Buddhism seems to be primarily an interpretation of Western Christians, who note how Buddhists treat such things as "gods"—subordinating them, regarding them as marginal, or even denying their existence. However, the notion of the universal inevitability of change concerns arising, abiding, and passing away, and therefore Buddhists assert the "emptiness" of any thing's ability to remain whole and intact; as we shall see, destruction has been a deep concern of Buddhist thinkers, not because of nihilism, but because of their desire to come to terms with it. Buddhist thought is "deconstructive," but Buddhists regard this deconstruction as the dissolving of delusions rather than the denial of meaning or existence. We hope that our study of destruction might contribute to the understanding of various factors resulting in religious and cultural violence and, if not to prevent it, at least to stimulate the identification of alternative, more subtle forms to manifest dissent and promote change.

Acknowledgments

Over all these years, we have become indebted to many people who have supported our work on this strange topic in many ways. Fabio would like to thank Patrizia Violi, the late Nino Forte, Motoyama Kōju, Fujita Ryūjō, Ishizuka Jun'ichi, and Yamaguchi Masao. Eric would like to thank Mark Jordon, Vivian Gold, Dorothy Reinders, Bill Gilders, Li Hong, José Cabezón, Zoe Strother, Barry Flood, Leslie Brubaker, and other members of the Iconoclasm Network. Together we would like to thank our editor at Continuum Books, Lalle Pursglove; an anonymous reviewer for Continuum; our friend and ally Richard Clay, who together with Stacy Boldrick organized the conference "Iconoclasm: Contested Objects and Contested Terms" at the Henry Moore Institute in Leeds in 2001 (the paper we presented at that conference—later included in Boldrick and Clay (eds), 2007—constitutes one of the pillars of this book); all friends, colleagues, and students who heard or read versions of portions of this book, for their precious suggestions and scathing criticism. Finally, we would like to dedicate this book to our mentors and friends, Allan Grapard and Bill Powell.

Earlier versions of portions of this book have appeared in the following publications:

Rambelli, Fabio (2001), "Seizō hakai, kigō hakai, seisei hakai: seinaru mono no hakai kōi ni kansuru ichikōsatsu," *Hikaku bunka ronsō*, 8, September, pp. 15–43.

— (2002), "Secret Buddhas (*Hibutsu*): The Limits of Buddhist Representation," *Monumenta Nipponica*, 57/3, Autumn, pp. 271–307.

Rambelli, Fabio and Reinders, Eric (2007), "What Does Iconoclasm Create? What Does Preservation Destroy? Some Observations from Case Studies of China and Japan," in Stacy Boldrick and Richard Clay (eds), *Iconoclasm: Contested Objects, Contested Terms*. Aldershot, Hampshire, England: Ashgate, pp. 15–34.

Reinders, Eric (2004), "Monkey Kings Make Havoc: Iconoclasm and Murder in the Chinese Cultural Revolution," *Religion*, 34/3, pp. 191–209.

— (2005), "Recycling Icons and Bodies in Chinese Anti-Buddhist Persecutions," Res: *Anthropology and Aesthetics*, 48, Autumn, pp. 71–8.

All previously published material, however, has been revised, rewritten, and expanded here.

The epigraph is from Ludwig Wittgenstein, *Culture and Value*. Edited by G. H. von Wright. Translated by Peter Winch. Chicago: University of Chicago Press, 1980, p. 3, and Oxford: Wiley-Blackwell, 1998, p. 3e, with permission from University of Chicago Press and John Wiley and Sons, Ltd. An excerpt from *Born Red: A Chronicle of the Cultural Revolution* by Gao Yuan, Stanford: Stanford University Press, 1987, appears with permission of Stanford University Press.

We thank the publishers for their kind permission to reprint copyrighted material in this book.

Part One

Stuff

Materiality and Fragility of Dharma

1

Buddhist Objects, Buddhist Bodies—An Outline

Buddhism has been a massive, imposing presence in East Asia. People were surrounded by concrete, material signs of Buddhism, by Buddhist objects, and also by "non-Buddhist" objects redefined by Buddhism. Material intervention on the landscape defined monastic complexes and national networks of monasteries, such as Mount Kōya, Mount Hiei, and other institutions in the Kyoto-Nara region in central Japan, or the "Five mountains and ten monasteries" system of imperial monasteries in China. Pilgrimage sites feature pious graffiti, beautifully carved on cliff faces or crudely painted on large rocks. Other marks of Buddhism are more modest, such as altars for worship, stone images (*sekibutsu* in Japanese), and simple pagoda-like structures. Today in Taiwan we can see "Namu Amituo-Fo" (Praise to Amitabha Buddha) bumper stickers on cars, doors, and telephone poles.

Of course, the Buddhist presence is not limited to objects and tokens of the sacred. A vast clergy, numbering hundred of thousands, has dedicated itself to spreading the teachings—an activity that involves the production, manipulation, and consumption of material objects, wealth, space, and nature. In typically hostile terms, Jacques Gernet summarized the situation thus: "There was an overabundant monastic community, incessant construction work, a prodigious consumption of metals, wood, and cloth" (Gernet, 1995, p. xv). People directly involved in activities related to Buddhism were not limited to officially ordained monks and nuns, but included novices, acolytes, itinerant religious specialists, diviners, performers and storytellers, artisans, merchants, laborers, and peasants more or less loosely associated with religious institutions. The bodies of most of these people were Buddhistically marked, as it were—ordered and/or ordained in ways that indicated their Buddhist affiliations, or at least the fact that their activities were related to Buddhism.

The reality of Buddhism for the majority of East Asians, who had neither the time nor training (nor necessarily the inclination) to engage in meditation or philosophy, was *visible,* as objects and wealth. Today, traces of the material presence, wealth, and labor associated with Buddhism—albeit minimal if compared with medieval or premodern eras—are still conspicuous in China and Japan. This is the stuff Buddhism is made of, and it would be unfair, if not misleading, to deal with the history of Buddhism only in a dematerialized and disembodied fashion, as if doctrines and rituals had no material and bodily substratum and import. Not surprisingly then, the maintenance, management, and control of such an array of objects and such large numbers of people have been among the major preoccupations of Buddhist institutions. At the same time, state institutions have been weary of the sheer autonomy, wealth, and mobilizing power of Buddhist institutions; whenever and wherever they had been powerful enough, states have relentlessly attempted to control the extension of the Buddhist material dimension—if needed, also through confiscation, physical violence, destruction, and murder. At times, attempts to restrain the power of Buddhist institutions became indictments of Buddhism per se.

Buddhists were painfully aware of the problems generated by their own wealth and success. Objects and bodies are essentially transient and prone to decay; objects break and people rebel. Wealth and power are signs of Buddhism's success, but at the same time they can arouse greed and the suspicion of state institutions. Buddhist thinkers stressed that Buddhist institutions' failure to control people and objects (including Buddhists' own failure to control themselves in their interactions with sacred objects and wealth) would eventually result in the end of Buddhism. Indeed, prophecies about the end of the Dharma (Ch. *mofa*, Jp. *mappō*) are inextricably intertwined with misuse of objects and clerical misbehavior.

In light of the above considerations, this chapter addresses a number of issues of East Asian Buddhism's materiality. Given the scope of the subject, we do not aim at an exhaustive account, but we limit ourselves to outline what we envision as the background for our investigations of religious destruction in East Asia. First, we discuss the importance of material objects for religious purposes; this involved the development of strategies to sacralize (i.e., transform in Buddhist ways) things and activities—including some that were originally secular and/or unrelated to Buddhism. Second, we deal with attempts to establish forms of embodiment of Buddhism, in particular the development of a clerical *habitus*—this amounting to the attempt to sacralize the human body. Third, we examine the widespread tendency in East Asian Buddhism to blur distinctions between animate and inanimate objects, icons, and bodies; particularly relevant in this context are the rituals of animation of sacred objects, but also the attempts to pattern monastic bodies upon an idealized Samgha (clergy) as based on monastic precepts (*vinaya*) and visual images of patriarchs. Fourth, we present instances of Buddhism's ambivalence toward materiality. While Buddhist institutions needed wealth and power in order to expand and carry out their salvific mission, excessive wealth and power in themselves were essentially against the basic tenets of Buddhism as a religion of renunciation. (It is worth mentioning that this Buddhist ambivalence predates religious studies scholars' own prejudices against materiality and preference for disembodied and immaterial "spirituality.") Fifth, we address the pervasive awareness of the fragility of the Dharma, expressed in forms that range from monastic rules to prevent wear-and-tear at temples to doctrines about the end of Buddhism. The end of Buddhism involved both the clergy's wrong attitude toward and misuse of objects, and secular attacks against religious institutions. In particular, the end of Buddhism was not simply the end of "belief" in Buddhist teachings, but was described in concrete terms as the destruction of Buddhist sacred objects and the dissolution (understood as both "misbehavior" and "disrobing") of the collective body of Buddhists. Finally, we indicate the need for a new conceptual apparatus to address in a systematic way religious violence in East Asia—destruction of sacred objects and violence against Buddhists, their conceptual motivations, and Buddhist responses to them. In this final section of the chapter, we introduce an expanded and modified notion of "iconoclasm" as a useful conceptual tool to deal with these complex issues.

We should make it clear from the beginning that by "sacred" we intend nothing ontological or theological about the existence of gods and other supernatural beings; we are methodologically agnostic. Culturally, though, sacredness is a category of separation, always in a binary relation with the "profane" or with the ordinary world. Chinese and Japanese temples, their objects and clerics, are defined by various pairs of terms, such as "outside the secular world (Jp. *shusse*, Ch. *chushi*)" vs "in the secular world (*seken*, *shijian*)"; "outside the household (*shukke*, *chujia*)" vs "in the household (*zaike*, *zaijia*)"; "sagely (*sei*, *sheng*)" vs "vulgar (or lay, *zoku*, *su*)"; or by possessive identification with entities such as buddhas, bodhisattvas, devas, spirits, *kami*, and other powerful personalities.

A look at Buddhism's material dimension

Ritual implements used at temples or at home include clerical robes and accessories (the three robes, staff, scepters), temple ornaments (banners, garlands, canopies, bells, boxes, lanterns, tabernacles), altar implements (vases, vessels, little *stūpas*, etc.), musical instruments (*biwa* lute, *waniguchi* and other cymbals, *mokugyo* drum), incense tools (incense burners, incense), *argha* (Jp. *aka*, consecrated water), ritual paraphernalia (*vajra* thunderbolts), reliquaries, rosaries, and so forth.[1] In addition, we find in the religious life of East Asian Buddhists many "unofficial" gadgets that defy classification, such as postcards, temple stamps, souvenirs, jewelry, bumper stickers, and temple literature (illustrated admission tickets, maps, pamphlets, books, journals, videos and DVDs, tapes and CDs, and homepages). In Taiwan, some temples even offer ashes from their incense burner for people to take home in small paper packets. Many of these objects and services have niche markets: clergy, children, young people, students, elders, men, women, professionals, heads of households, and tourists. An interesting function of these portable, purchasable objects is that they can be used to sanctify tokens of modern life (cars, stores), giving them the mark of Buddhism. Objects are related in a way or another to some of the most important moments in life: birth, death, study, beginning a career, finding a partner, and family life. When considering the status of sacred objects in Buddhist traditions, it is interesting to note that Dharma transmission itself was often envisioned as a form of sacred transaction involving the exchange of gifts between master and disciple, especially the former's robe and bowl.[2] Buddhist notions of transmission were strongly influenced by Chinese ancestral practice, and in turn the Dharma transmission became a model for succession in general within the Buddhist world: a master transmitted to his disciples not only the Dharma but also his own possessions.

The earliest accounts of the arrival of Buddhism to Japan agree on the items that constituted the transmission of Buddhism: scriptures (no title is given), a Buddha image, and implements used to deal with the image ritually—in particular, bathing the statue (Rambelli, 2010b, pp. 148–9). Similarly, the mythic introduction of Buddhism to China, in which the Han Emperor Ming (28–75 CE) dreamed of a "golden man," resulted in his sending envoys West, who returned with a Buddha *image*. These accounts of the arrival of Buddhism to Japan and China, confirmed and expanded by countless subsequent narratives, emphasize that the Dharma, in fact, is embodied by objects (and bodies), and the spread of such objects was the spread of the Dharma. Japanese priests who went to China to study Buddhism prepared long and detailed accounts of the things they brought back from their perilous voyages. The "quest for the Dharma" (Jp. *guhō*) was also a quest for *tokens* of the Dharma such as scriptures, images, relics, and ritual implements. The "Catalog of Imported Items" by the Japanese monk

Kūkai (774–835) is particularly representative examples of this genre. Among "Buddha images" (*butsuzō*) he lists mandalas and portraits of patriarchs. He also lists ritual implements (*vajras*, etc.), relics, other images, shell trumpet (*horagai*), mendicant bowl, *kāsāya* monastic robe, and ceremonial bowls for offerings with precious stones (Hakeda, 1972, pp. 140–50).

East Asian Buddhists did not only produce, circulate, and collect sacred objects, they also theorized about them and about materiality in general, in what are some of the most fascinating instances of Buddhist thought. Chinese and Japanese exegetes developed a series of concepts that refer to the realm of the inanimate objects or nonsentients, that is, material objects and entities devoid of a conscious mind, which constitute and furnish the material space where sentient beings living in the Six Destinations (Jp. *rokudō* or *rokushu*, Ch. *liudao* or *liuqu*) and buddhas live and operate. In particular, "nonsentients" (Jp. *hijō* or *mujō*, Ch. *feiqing* or *wuqing*), "realm of objects" (Jp. *kikai*, *kisekai* or *kiseken*; Ch. *qijie*, *qishijie*, or *qishijian*), and the material environment (Jp. *ehō*, Ch. *yibao*, lit. "karmic support"), refer to the Umwelt of buddhas and living beings, that is, the environment and the material living conditions in which beings find themselves as a consequence of karmic retribution.[3] In Japan, these terms referring to materiality and the environment were considered synonymous with more concrete expressions such as "plants and the territory" (*sōmoku kokudo*), "plants, rivers, bricks, and stones" (*sōmoku kasen gareki*), or more simply "plants" (*sōmoku*).

In more specific terms, Buddhism classified sacred objects on the basis of the modality in which the sacred is produced or manifested in them. Thus, we find alternate bodies (Jp. *funjin*, lit. "separate" bodies), condensations of the cosmos (mandalas), sacred receptacles, and objects sacralized through immolation. Alternate bodies are doubles of the deities, sort of fractal reproductions of the original sacred entity, in which the totality is identical to its parts or fragments; among this category of objects are relics (Jp. *shari*), but also talismans and amulets (*o-mamori* and *o-fuda*). Relics are parts of the body of a Buddha, a saint, or a past master, but amulets and talismans must be "charged" with the "spirit" of the deities of which they are alternate bodies. Mandalas are defined as condensations of the totality of the cosmos; trade tools of secular professions (such as the merchants' scale, *hakari*) and ritual implements (such as the *vajra* club) were understood as mandalas. Sacred receptacles are objects whose status is situated somewhere between the two previous categories; as examples we find the family Buddha altar (*butsudan*) and the funerary tablet (*ihai*). The family altar, modeled after the cosmic mountain Mount Sumeru and playing the function of a feretory and, ultimately, of a temple, is in itself a cosmological model but not necessarily a mandala (since it is not coextensive with the actual form of the cosmic Buddha). The funerary tablet (doubled into *ihai* kept at home and *sotoba* kept at the family temple) is modeled after the *stūpa*, which is both a cosmological model *and* a mandala, as in the five-element mandala (*gorin mandara*), but it contains the spirit of the ancestors, rather than *being* the ancestors (Rambelli, 2010a). Finally, objects sacralized through immolation include old objects for which memorial services (*kuyō*) are held; here, these objects function as sacrificial offerings: while all of the objects mentioned above are sacralized after construction by summoning within them the presence of the buddhas (and in this sense, they are equivalent to buddha images), everyday objects are sacralized at the end of their life before they are disposed of, and in this sense they function a little as dead ancestors (see also Rambelli, 2007).

What are the doctrinal bases of this interest in the material dimension of reality? One of the most significant intellectual tendencies in Mahayana is the increasingly materialistic interpretation of the concept of *dharmakāya* (Ch. *fashen*, Jp. *hosshin*). From an originally abstract, purely conceptual and non-experiential entity signifying the fundamental nature of the Buddha and his teachings, it came to be gradually envisioned as the overall substratum of the Buddhist cosmos to the point that, in Japanese esoteric Buddhism, the cosmos (Sk. *dharmadhātu*, Jp. *hokkai*, Ch. *fajie*) was identified with the supreme Buddha-body (*dharmakāya*); this was also related to the development of the discourse on original enlightenment (*hongaku*). This latter move, clearly indebted to Indian Brahmanical cosmology and ontology, became the definitive justification of Buddhist interventions in the secular, material realm. However, by then many important cases of Buddhist monumentality had already been created. An influential form of intellectual substratum for Buddhist materiality, from monumental interventions to the production of countless objects, can be found in Avatamsaka (Ch. Huayan, Jp. Kegon) thought, with its positing of a universal Buddha, Vairocana, and in its emphasis on the unobstructed interpenetration of abstract principles (Ch. *li*, Jp. *ri*) and phenomena (*shi*, *ji*) on the one hand, and interpenetration among phenomena (respectively, *lishi wuai*/*riji muge* and *shishi wuai*/*jijimuge*) on the other. According to this vision, the cosmic Buddha and its principles were present in each single particle of dust of the universe, and each single particle was an essential component of the whole. In this respect, contrary to received interpretation of Huayan/Kegon thought as overly abstract and speculative, we would like to suggest that certain cases of extensive material interventions on the landscape and the everyday life of the people, from monumentality to pervasive diffusion of traces of the presence of the Buddha in the everyday world, were the results of attempts by Buddhist creators to give shape to the Avatamsaka worldview.[4]

Buddhist manifold interventions in the realm of materiality were also the result of devotional practices aimed at the diffusion of Buddhism through proliferation of sacred objects (temples, stupas, icons, texts, sundry implements) on the one hand, and construction works (bridges, roads, hospitals, land reclamation, irrigation works, wells) for public benefit on the other. Some of the most ancient Buddhist scriptures, including the *Āgamas*, extol the merit-making value of these construction and manufacturing activities.[5] Scholars have generally focused on the production of temples and images as primary merit-making activities, but in fact scriptures and the actual praxis of Buddhist institutions show that there was no clear distinction between sacred and secular works.[6] As mentioned by Janet Goodwin, documents by entrepreneurial, fundraising monks (*kanjin hijiri*) from medieval Japan about construction activities they promoted are written in a language "full of physical metaphors suggesting the equivalence of building temples and saving souls. The equation was the same in other cases, even when the construction project was a bridge" (Goodwin, 1994, pp. 140–1). We should keep in mind that merit-making is a diffusive endeavor, affecting promoters, executors, and beneficiaries of the construction. Another devotional aspect of Buddhist monumentality and landscape interventions might be related to visualization practices to the Buddha, especially those Buddhas endowed with a cosmic nature that makes them hard to experience and to relate to. Thus, decorated caves at Dunhuang and elsewhere provided a sumptuous description of the Buddha-land, and gigantic statues of buddhas were powerful aids to visualize their glory and cosmic dimension. In addition, extensive temple complexes, sometimes hosting several thousands monks, were clear reminders of the prosperity and power of Buddhism in this world.

In many cases, Buddhist materiality resulted in the accumulation of enormous wealth and mobilization of vast manpower. We should not forget that large Buddhist religious institutions were research institutions (dedicated to Buddhist studies and the sciences—mathematics, astronomy, medicine, architecture), and providers of welfare services. In particular, they ran hospitals, public baths, orphanages, and public education facilities; they organized food and clothing donations, and fundraising activities for secular construction works (dams, land reclamation, bridges, roads, etc.). Many professional groups were affiliated to them, such as merchants, carpenters, artists, and performers. Due to the number of people directly or indirectly affiliated with them and the range of the activities they supported, Buddhist temples were important economic centers as well. Their activities were not necessarily conspicuous consumption in which resources were wasted (as according to Confucian and secularist critiques); on the contrary, the investments did produce significant economic returns, not only for the religious institutions involved but also for many people related to them, either as wages or as what we would call today welfare benefits. Thus, wealth and prestige coming with it came to be closely associated with soteriology, because they enabled religious institutions to carry out numerous Buddhist activities (study, education, proselytizing, charity).

For anti-Buddhists, "Buddhism" was visible as waste and scandal, a drain on social resources. As Reischauer has written, "throughout Chinese history the chief cause for overt opposition to Buddhism has been economic rather than religious" (Reischauer (trans.), 1955, p. 218). Buddhist materiality was an easy target, because of its visibility, diversity, and quantity. Indeed, in China, the most effective philosophical attacks on Buddhism (during the Song revival of Confucianism) came after the most effective economic attacks, especially the Huichang persecution in 845 (Reinders, 2004b, 2005). In Japan, the destruction of the major centers of monastic power by Oda Nobunaga and Toyotomi Hideyoshi in the late sixteenth century was followed by a sort of nationalization of Buddhism by the Tokugawa regime as an appropriation by the state of the symbolic and social capital of religious institutions; but the resources spent on the extensive network of temples and priests became the target of widespread anti-Buddhist discourses, often grounded on economic reasons, especially in the eighteenth and nineteenth centuries. The late Tokugawa and early Meiji anti-Buddhist persecutions were aimed primarily at a redistribution of the various forms of Buddhist capital (see Chapter 2).

We should note here that the close association of wealth with the soteriology characteristic of Buddhism poses an intellectual challenge: when secular authorities persecuted Buddhist institutions to appropriate their material possessions, did they simply target the profane, material aspects of Buddhism? Or rather were they questioning—be that in implicit ways—Buddhist soteriology as a whole? Indeed, can materiality be so easily separated from spirituality?

In recent years, the study of material culture has become more frequent and normal in religious studies,[7] which now frequently address "such questions as how objects are experienced, what needs other than functional ones they answer, what mental structures are interwoven with—and contradict—their functional structures, and what cultural, infracultural or transcultural system underpins their directly experienced everydayness" (Baudrillard, 1996, p. 4). If we pay attention to Buddhist practices, we see that material objects are extremely significant, and not only in their practices but also in their ideas. Material objects of many kinds are essential parts of Buddhist practice as a necessary part of what constituted "being a Buddhist." Indeed, there is very little meditation, doctrinal study or philosophical debate in East Asian Buddhist households; in a publicly visible sense, their Buddhist identity is usually associated with ownership of Buddhist objects (such as the home shrine) and visits to Buddhist temples. In Japan, the Tokugawa system whereby households registered with Buddhist sects has made this association particularly strong, but the general point holds true of China and Taiwan. Again, this should not be understood as a form of "degeneration." As Jean Baudrillard showed, objects establish a "more or less consistent system of meanings" and behaviors, and play an important role in establishing and representing their users' identity and world-view (ibid.). Thus, the assumption of a mutually exclusive opposition of the spiritual and material would be misleading. Spiritual experience can be simultaneous to interactions with material objects. The object of metal around the neck of a Buddhist is the specific material focus of that person's identity as a Buddhist.

Crafting Buddhist bodies

Buddhist intervention in the world was not limited to nature and material objects. The body of practitioners, especially that of members of the clergy (*samgha*), was the subject of numerous and detailed sacralization strategies. Sacralization of the body was at times described with analogies to the production of objects.

One of the earliest Buddhist texts to circulate in Chinese makes the analogy of a monk and a manufactured object explicit: "Practicing the Way is like forging iron: if you gradually but thoroughly cast out impurities, the vessel is sure to come out well" (translated in Sharf, (trans.), 1996a, p. 370). The term *duantie* (forging iron) and related terms such as *duanlian* (refining) had both literal and metaphoric uses, referring to self-cultivation. The term *chengqi* (making a vessel) is from the classic *Liji* (Book of Rites), though it seems to have had only literal meanings there. The famous Vinaya Master Daoxuan (597–667), best known for his commentaries on monastic discipline (*vinaya*) and for his activism in defending monastic privileges from various political attacks, wrote: "The body is like wood and stone; if it is engraved and polished, it successfully makes [one into] a vessel" (of the Dharma-body).[8] This metaphor of the monastic body as the product of the artisan's skillful labor bears striking resemblance to the production of an icon. The pre-monastic body is raw material, subjected to skilled labor in the production of an object of obeisance. The produced nature of the monastic body is clear in Daoxuan's correlation of the manufactured Buddha image and the monk's body: "metal, stone, mud, and such raw materials can [be made to] display the appearance of the True Image [of the Buddha]. The Dharma-clothes and cut hair determine the form of the completed monk."[9] Daoxuan distinguished between monks and laity in their acts of manufacturing Buddhism:

> The laity use metal, stone, earth, wood, ivory, horn, cotton, and silk to make a Buddha image. The ordained cultivate the five-part [attributes of the] Dharmakāya (*wufenfashen*), study the Three Buddhist Practices, and this is called making an image (*zaoxiang*). (It is said that the laity use phenomenal things [*shi*] to make things, whereas the ordained make things with the Dharma.) The laity use paper and plain silk, bamboo and silk, pen and ink for writing to make sutra scrolls. The ordained use hearing, thinking, and cultivating wisdom [the three modes of attaining wisdom] to make the Dharma (*zaofa*). The laity uses grass, wood, walls and roofs to make a temple. The ordained use the palace of bodhi-nirvana-wisdom and the Mahayana dwelling where the ten thousand practices reside as a temple.[10]

The laity make Buddhism with things; the Samgha make themselves Buddhas. Thus, Daoxuan envisioned the production of the three treasures in terms of physical objects on the one hand and as practices of self-creation on the other hand. Whereas laypeople take some raw material and make physical things, monastics produce a Buddha image by applying cultivation techniques to themselves, they "make dharma" by transforming their own consciousness, and they live inside a nexus of transcendent wisdom and cultivation practices. The products of the laity are obviously perishable, whereas the eternal realities of the Buddha are not.

At least since Foucault, it has become commonplace to consider the "body" as a produced or constructed object. But to understand the iconization of the human body, and to better identify the target of violence, it is useful to introduce a more precise term, *habitus*. Pierre Bourdieu used the term to name our embodied knowledge, our bodies made different, not randomly but in systems of "dispositions."[11] *Habitus* appears as a style of behavior, marking the difference between classes or genders or between social groups. If acted, it is artificial (a product of artifice) until such time as, through repetition and imitation, it becomes "natural" to a body, un-self-conscious in its stylization and authoritative in its given field of power. It is "a feel for the game," and the "air of a soldier" as discussed by Michel Foucault:

> By the late eighteenth century, the soldier has become something that can be made; out of formless clay, an inapt body, the machine required can be constructed; posture is gradually corrected; a calculated constraint runs slowly through each part of the body, mastering it, making it pliable, ready at all times, turning silently into the automatism of habit; in short, one has "got rid of the peasant" and given him "the air of a soldier." (1979, p. 135)

Throughout *Discipline and Punish*, Foucault used a variety of terms to suggest what he called the "microphysics of power" (ibid., p. 28) or the "political technology of the body" (ibid., p. 26): these are "humble modalities" (ibid., p. 170), operating "at the level of the mechanism itself" (ibid., p. 139) and measuring the "smallest fragment of life" (ibid., p. 141), so that in the end, we can see how "[d]iscipline is a political anatomy of detail" (ibid., p. 139). We will see below some of the rules written for use in training new arrivals to monasteries. Such texts are evidence of meticulous regimes governed by written and oral "guides to practice," creating a micro-discipline of the body, precisely, a habitus. The result of disciplinary practices is a transformation of the self, a subject-making widely accepted and eventually un-remarked on. Hence, "habitus consists of a set of historical relations 'deposited' within individual bodies in the form of mental and corporeal schemata of perception, appreciation, and action" (Bourdieu and Wacquant, 1992, p. 16). These deposits are visible as manner, style, and body display.

The utility of the distinction between the body and the habitus is clear when we consider the ordination rituals, the distinct clothing and hairstyle, the submission to Vinaya discipline, the whole construction of the Samgha—these things do not create a "body" in the literal sense: *habitus* is a more precise term for what bodies can become. Texts on obeisance, for example, emphasize repeatedly that a pious layman should bow to all monks, even those of dubious morality. Once the body of a monk has been forged by Vinaya micropractices, the *habitus* of the Samgha—made visible by physical features (the shaven head), objects (the clerical robe), and deportment—put all its members into a separate category. The ontological rhetoric should not obscure the fact that the *habitus* required concentrated labor to be created and maintained, namely the endless process of training and re-training bodies, as well as the construction of ritual sites and the dissemination of appropriate images of transcendence. A monk was an image of Buddhism, but Buddhism also represented him.

In what ways were Buddhist bodies—of male and female novices, nuns, and monks—crafted as sacred? What made the distinction between Samgha and laity a matter of the body? The ordination ceremony relied on symbolism of dying to the world, including renaming, and throwing away worldly wealth. The cutting of the ordinand's hair re-enacted the Buddha's renunciation after he fled his palace. The moment that the knife cut the first hair of the topknot, the ordinand's body became identical to that of the Buddha.[12] *Vinaya* literature, with its long list of rules, is evidence of meticulous regimes governed by written and oral "guides to practice," with an emphasis on overcoming bad physical habits, creating a micro-discipline of the body, a *habitus* as the basis for the cultivation of a transformed mind. Ultimately, in principle, "the Buddha's Dharma takes mind as the root, and body as the branches."[13] But for Daoxuan, the training of the body has a practical priority: "first you make your body comply and then in the end you can remove the mind's delusions."[14] Through correcting the exterior form, the individual's buddha-nature is manifested. Daoxuan makes a similar point about the material Buddha image: when made correctly, "it is the vessel (*qi*) of the Dharma-body."[15] By correcting the exterior form, the individual's buddha-nature is manifested.

The notion of the monk "being" or "becoming" a Buddha can be treated in various ways. One prevailing assumption was that in the correct internal and external recreation of the Buddha's body, speech, and mind, an abbot was, de facto, enlightened, so that in speaking a Dharma talk from the high seat, he was expressing his Buddha-nature, channeling (as it were) the best part of him. However, becoming a Buddha was never a purely inner accomplishment in China:

> When one turns to descriptions of Buddhist "meditation" found in traditional *dhyāna* manuals, however, one discovers that terms such as *dhyana*, *yoga*, *samatha* (concentration), *vipasyana* (liberative insight), and *samadhi* (absorption) refer not simply to states of mind but to highly formalized procedures in which all aspects of a practitioner's physical regimen, behavior, and deportment are prescribed in exacting detail. (Sharf, 2005, p. 260)

One of the important developments in recent studies of Buddhism has been an effort to conceive of "becoming a Buddha in this very body" as a *performative* process, in which social position and ritual practices transform the masterful practitioner's body into an "image" of the Buddha-body.

Indeed, Daoxuan went to great lengths to maintain the properly crafted monk's body. He wrote various instructions and essays related to the objects of the monk: his robe, his bowl, the traditional list of objects allowed by the Buddha as personal property of monks, including the three robes, the bowl, a sewing kit, and a water strainer. Yet it might be said that these properties were only "props" for Daoxuan, and the real target for his craft was the monk's body, an impressive body characterized most of all by *weiyi* (Jp. *igi*, Sk. *prabhava*)—a term variously translatable as dignified demeanor, respect-inspiring decorum, or good manners.

In what sense was a monk an "image"? At least in the sense of reputation, public perceptions, and matters of "public relations," each monk's body was a visible local representative of the Three Treasures, which was always discursively vast—an entity or category spanning time and space in ways the individual monks never could—an immortal institution, a timeless truth, and an omniscient consciousness. But as something that could be seen, the monk's body was what Daoxuan called the "tangible Samgha" (see below).

In hagiographic literature, the idealized death drama of famous monks has them meet the destruction of their own body with supreme decorum. Typically, the monk enters a state of *samādhi* just prior to death, just as part of the consecration ritual involved placing the Buddha(-image) in *samādhi*. Hence, as the monk's body changed from living to dead, from a sentient being to an object, he resembled a consecrated image—an object of continued obeisance, only apparently inanimate, bodily transformed into a miraculous relic (*śarira*). In a few cases, the monk's body was preserved in its seated form rather than cremated, and enshrined as a "flesh-body" (*routi*) on an altar. Bodily holiness was assumed to be manifest in unusual and/or imperishable bodies, and holy monks were said to stay fresh and even fragrant after death. Some of these holy bodies were even lacquered—literally made into statues—which completed the parallelism (Robson, 2011; Faure, 1996, pp. 169–73). Relics—the organic matter of particularly holy monks—were placed inside images as a further mechanism of animation. As Bernard Faure writes,

> The icon is also a kind of stupa or mausoleum, just as the stupa is an architectural body, a (living) double of the dead person. Like a mummy, a relic, or a devotee temporarily sunken into *samādhi*, it is alive, present, awake. (Faure, 1996, pp. 252–3)

In still other ways, Buddhist rhetoric asserted the sacrality of the monk's body. The emphasis in Buddhist (especially Chan and Zen) meditation on absolute stillness of mind and body, of "just sitting" in a lotus position, might be seen as an attempt to become a statue—to "become a Buddha [-image] in this very body" (Jp. *sokushin jōbutsu* [*-zō*], Ch. *jishen chengfo* [*-xiang*]). The well-known assertion of "Buddhahood in this very body" may refer to the immediacy of "sudden" enlightenment, but it also indicates a sacralization of the monk's body effected through ritual. The sacralization of monks' bodies in Chan/Zen involved a transmission of the Buddha's enlightened mind; this transmission was performed with both body and mind. The emphasis on the direct continuity of enlightenment from the Buddha to his many disciples again negated or at least demoted the distinctions between them.

In these and other ways, then, the monastic body was presented as something else, and, at least in certain circumstances, treated as if it were the Buddha:

> The Chan abbot . . . literally took the place of a consecrated buddha icon and accepted the offerings and worship of the supplicant(s). Participants in such rites approached the abbot as if they were coming face-to-face with a living buddha. The icon of wood, stone, or metal has been replaced with a living icon of flesh and blood. (Sharf, 2005, p. 261)

It should be noted that the Buddhist transformation of the body was not limited to the ordained clergy, but affected to various degrees all practitioners, in meditation, ritual, pilgrimage, and various practices. Such transformation normally involved marks (clothes, apparel, haircut) and objects (the pilgrim's staff and robe). In addition, clerical bodies in particular were fashioned also based on visual models such as statues and paintings. Thus, we can see a close relationship between sacred objects and sacralized bodies in Buddhism that requires further explanation. Finally, the mirroring of icons and clerical bodies occurred also in cases of anti-Buddhist violence, which targeted humanoid features of icons (head, eyes, hands) and the iconic aspects of the human body (through disrobing, disfiguring, humiliation).

Buddhist rhetoric of animation

The mirroring of animate and inanimate did not occur only between icons and human bodies, but seems to constitute a more pervasive phenomenon. In fact, one of the most fascinating features of East Asian Buddhism is the blurring of distinctions between sentient beings and inanimate objects, as best represented in doctrines about the sentiency of natural objects such as trees and stones. Premodern doctrinal discussions about the status and the role of the inanimate world (natural environment and human-made objects)—especially ideas that plants and rocks are endowed with Buddha-nature—attempted to give ontological, cosmological, and epistemological grounds to religious objects. These doctrines may have been related to pre-Buddhist, animistic outlooks in India and in East Asia, but as developed by Buddhist authors, they came to be associated with rituals of animation, in which specific objects were infused with some form of spiritual entity (often not very well defined, even in doctrinal treatises). Such animated objects formed a material continuum ranging from sacred icons (relics and statues of buddhas and bodhisattvas) and ritual paraphernalia (such as the Japanese Buddhist family altar or *butsudan*) to secular professional tools (the merchant's scale, the samurai's sword). According to some versions by Japanese esoteric Buddhist authors, all of these objects are transformations of the Buddha-body; all are involved in some kind of salvific activity. The difference between icons (and stupas) and other objects sometimes lies in the explicit presence of a "sentient principle" (mind)—which is potential or hidden in inanimate things, but actualized, localized, and manifested in statues after the performance of a consecration ritual (see Rambelli, 2007, pp. 43–4, 49). In general, objects owned by the Samgha were thereby owned by the Buddha, and shared in the Buddha's otherness and sacrality. The proliferation of objects endowed with salvific power helped bridge the fundamental difference between the universal, formless Dharma and its particular transmission or instantiation—if everything touched by the Buddha is endowed with salvific power, but certain objects are better foci of that power, then the more you have the better. The sense that an icon was imbued with a transcendent presence was a responsibility, and also a reason to create and ritualize more icons.

The modern Japanese scholar Shimizu Tadashi makes an explicit connection between the use of sacred objects and salvation in Buddhism. He writes: "Buddhist sacred objects (*butsugu*) are tools for the worship (*kuyō*) of Buddhas . . . and ancestors . . . Buddhists believe that through worship one becomes a Buddha" (Shimizu, 1999, p. 7). The notion of "becoming a Buddha" through "just sitting" in meditation is well-known, but devotional rituals were also conceived as means of embodying the Buddha. This immanence is possible because these religious spaces, movements, and objects are different from everyday things. As a consequence, the use of sacred implements generates a sensibility that is different from that of everyday activities. Shimizu adds: "Buddhist objects are organically related to rituals; rituals reflect the Buddhist's life (understood as Buddha's path [*butsudō*]); and the life of a Buddhist is centered on the clergy (*Samgha*)" (ibid.). Here Shimizu refers to connections between the Buddha, sacred objects, and the clergy dating back to the origin of the Mahayana tradition in India. He argues that the use of sacred objects in ritual is an extension of such ancient Indian cults that reflected the influence of Buddhist spirituality in laypeople's religiosity (ibid., p. 8). In other words, sacred objects are embodiments of the spirit of lay Buddhism. Shimizu suggests that the development of cult centers determined the proliferation of sacred objects, already classified by the Vinaya into three main categories: "temple's implements" (*garangu*), "initiation implements" (*kanjōgu*), and "deportment accessories" (*igigu*). These were related, respectively, to the sacred place were ceremonies took place, to ceremonies themselves, and to the clergy and ritual specialists (ibid., p. 14).

We find in Shimizu a distant echo of Jean Baudrillard who, in his classic study of the impact of objects on the life of people, wrote that objects are "reflection[s] of a total order." He added: "Beyond their practical function . . . objects . . . have a primordial function as vessels, a function that belongs to the register of the imaginary," in the sense that they are the "reflection of a whole view of the world" which they help instantiate and materialize (Baudrillard, 1996, pp. 27–8).

Now, idolatry—the (wrong) idea that objects are alive—has always been part of Western representations of various Others, whether in Biblical polemics against indigenous cults, in Protestant polemics against Catholics, in the condemnations of heathenism by missionaries, or in various theories of fetishes and animism. For the modern West, the Reformation was most influential in setting the tone for representations of "idolatry," and it was closely followed by a world-wide expansion of European power and knowledge, which brought to Western audiences a vast array of supposedly living objects.

Buddhists in Asia have not been exempt from accusations of idolatry; indeed, their religious rituals explicitly attest to the real presence in their icons of Buddhas, bodhisattvas, or deities. As we will see more in detail in Chapter 3, Western discourses of idolatry affected in significant ways the development of religion (including iconoclastic instances) in modern East Asia, particularly in China where Christian missionaries were active. Also because of Western attitudes, the notion of divine presence in icons was somehow forgotten or edited out of modern Buddhism in the twentieth century, especially in its Western forms. As Bernard Faure remarked, "The notion of animated Buddhist icons has been repressed as a result of the modern and Western values of aestheticization, desacralization, and secularization" (Faure, 1998, p. 769). Asian and Western apologists "cleaned up" Buddhism, not only for Western inquirers who were frequently hostile to conventional Christianity, but also as part of an Asian re-invention of its own religious culture as more "scientific" and modern/Western. In the West, a turn toward Buddhism has been frequently also a turn away from the commercial, bureaucratic, ritualistic, and materialistic aspects of "organized religion." Despite the residual objects and rituals in American

Zen centers, for example, "Zen" is often presented as being not about objects, rituals, beliefs, or doctrines, sometimes not even about Buddhism, but about the unmediated experience of Being itself. In modern Buddhist apologetic writing, and to a lesser extent in academic writing, Buddhism has been pervasively dematerialized, virtually defined as philosophy and/or meditation. One finds many Buddha images in America today, but they tend to be decorative (in homes, gardens, and museums) and/or associated with specific ethnic cultures (in Thai or Chinese temples and restaurants, for example).

Under such circumstances, to re-acquaint Western readers with the role of icons in premodern Buddhism without also lapsing into a rhetoric of idolatry, and without implying an evolutionary model of progress in religion, requires careful control of language. At this point let us focus specifically on Buddha images, as they enjoy particular status as foci of ritual and devotional practices—a status that is perhaps related to their being mimetical re-presentations of the Buddha. What *is* a Buddha image—the "Buddha" who can be formed out of clay, touched, worshiped, but also knocked over and shattered? How does one reconcile the claims of Buddha's vast power with the fragility of his material substitute? East Asian Buddhists had available to them a sophisticated justification for the production of icons, which placed the icon at the heart of their project of salvation. In this case, as in most others, the primary rhetoric about religious icons, whether asserted or rejected, has been based on an ontology of presence, that is, the notion that there is a god or spirit or persona "in" the object.

Views of images as idols, as symbols, and as art, are inadequate explanations of Buddha images in Asia. Recent scholarship has come a long way to enrich and complicate our present understandings of Buddha images, by focusing on symbolic and ritual production, how they were created and how they were maintained. No longer merely a "symbol," the Buddha image now appears in much scholarly literature as an "object of power" (Brown, 1998, p. 45).

Images were imagined and treated as organic beings, not just chunks of wood or "bricks or tiles" (Gjertson, 1989, p. 264). "True" images have consciousness, feelings, will; they are active, animated, even though in a meditative trance-state. Overlaid with organic metaphors, they demonstrate an interplay among monks' bodies, Buddha images, and Buddha's body. Even for the laity, there was a series of correspondences between body parts of images and of devotees. Images needed clothing, such as elaborate robes and scarves. Images were also tuned in to the economic fortunes of the nation, and could make cryptic prophecies about political and military events (see, for instance, Grapard, 2003). In Buddhist literature of Medieval China and Japan, personal consciousness and agency were routinely attributed to images. Agency was attributed to icons firstly in their ability to move around, by walking or even flying. In Luoyang, "every night the image would walk around its base, leaving sunken footprints on the ground" (Yang, 1984, p. 208; on flying images see also Reischauer (trans.), 1955, p. 49). But the mobility of images was also of a non-miraculous kind, as when images traveled, in palanquins or other vehicles, just as rulers on tour. Images were also apparently aware of their own economic value. They could speak or shout, and apparently objected to the violation of their own status as owned capital:

> when some robbers attempted to steal the golden statue, the statue, joined by the bodhisattva statues, shouted at them in chorus. Alarmed and frightened, the robbers fell to the ground instantly. Having heard the calls of the statues, the monks came out and caught the robbers. (Yang, 1984, p. 56)

The rhetoric of animation also affected the way icons were made. They were created under strict guidelines, with attention to details of color, proportion, gesture, and attendant symbols. In case of wooden images, not every log would do; sculptors looked for what they called "spiritual trees" (Jp. *reiboku*) or "divine trees" (*shinboku*), in which they thought they could find traces of the sacred, or even the presence of a deity. Such trees were characterized by their peculiar forms, or by accidents such as having been struck by lightning, or miraculous occurrences. The East Asian Buddhist literature is rife with stories about trees and logs calling out to be sculpted into images, or putting forth light or perfume—as to draw attention to their miraculous power. Next, rituals were performed before cutting the chosen tree, and the sculptor had to undergo a regime of spiritual asceticism while making the statue. But nowhere is animation of icons clearer than in the "eye-dotting" or "eye-opening" ceremony (Ch. *dianyan, kaiyan*; Jp. *kaigen*). The ritual of "opening the eyes" is of Indian origin, and came to East Asia in esoteric texts of the Tang (Strickmann, 1996, pp. 165–211). The eyes of consecrated icons are covered during building renovations and relocation, and more finally when the image is de-consecrated. In fact, when the power of icons needs to be narcotized, as it were (such as during restoration work or in museum exhibitions), Japanese monks perform a ritual extraction of the icon's soul (*hakkenshiki*).[16] Moreover, relics, scriptures, talismans, and other objects were put inside icons as part of the effort to animate them. The ritual of consecration simultaneously creates a living presence and yet explains its apparent lack of animation. The Buddha is now "seated" or hypostasized in a body, but before he can behave as other bodies (moving, talking, eating), he is placed in a deep meditative trance, "a state of *samādhi*, a quiet catalepsy" (Faure, 1996, p. 252). No longer a mere material object, and not yet an active sacred presence, the Buddha image is immersed in meditation, much like the future Buddha Maitreya. Analogous rituals of consecration take place also in the inauguration of sacred places, buildings, and in a monk's ordination. However, whereas in the eye-opening ceremony, the spirit of a Buddha is infused in the image, in the case of buildings and spaces the site is proclaimed under the control of Buddhism and emptied from evil influences, so as to allow buddhas and their retinues to inhabit it.

Thus, images are anything but inanimate objects: they are very animated. They fly through the air. They appear in dreams. They walk, talk, shout, sweat, grow hair, and radiate light. They need clothing and demand proper courtesy. In short, they live. Strange or wonderful displays of the life of icons could certainly teach Buddhists about the power of the Buddhas and their images, but icons were not primarily didactic. They were thought of as living. Recent scholarship has indicated that even today in China and Japan some people at least attribute a living dimension to Buddhist icons (Horton, 2007).

We should say, however, that excessive focus on the "living" nature of Buddhist images might result in simplification. Buddhism has a long history of conflicting interpretations on the nature and function of sacred images, which have been variously characterized as ranging from pointers to a higher reality, iconographic repertories, ritual foci, all the way to animated entities (including the idea of "doubles" of their spiritual models). Focus on animation, as it happens in much recent scholarly literature on the subject, reduces the intellectual and ritual richness of the Buddhist tradition. Moreover, the fact that people talk about icons in (almost) human terms need not be taken literally; in fact, humanization is a common rhetorical strategy (it applies to all sorts of objects, from toys to cars to the stock market), and does not necessarily imply animistic beliefs on the part of the speaker. However, when ritual uses of images go beyond rhetorical emphasis, such as in cases or ritual humiliation or mistreatment of icons (as described among others in Ishizuka, 1993), it is necessary to distinguish between form of animism and fetishism—and we believe that the latter might be more important in the interactions with sacred images than we usually think. Furthermore, such rhetorical and ritual apparatus treating images as living entities could be used to emphasize the continuity (often bordering on identity) between icons and the bodies of monks and nuns. On the other hand, a supposed animation of icons was also a strong incentive for iconoclasts, who could stress with their actions that icons were in fact merely material, and powerless; in a further twist, Buddhists at times were able to claim that natural calamities that followed iconoclastic acts were in fact the angry responses of the damaged icons.

Recent works on icons in art history and the anthropology of art have attempted to consider the agency and power of artificial persona created by representation. One argument can be made for a fundamental sense of all images having "life," as outlined by David Freedberg. Freedberg argues that by our very recognition that an image is *of someone*, our cognitive processes necessarily bring images to life:

> Almost every image provides its beholders with clues to the organic presences registered upon it . . . The smallest number of clues suffices to precipitate the search for more. Response to all images, and not only ones perceived as being more or less realistic, is predicated on the progressive reconstruction of material object as living. (1989, p. 245)

Hence he asserts: "the time has come to acknowledge the possibility that our responses to images may be of the same order as our responses to reality" (ibid., p. 438). Such a universal human capacity to respond to images helps to normalize image consecration, iconoclasm, and many other phenomena. Indeed, Freedberg scorns our sophisticated methods of insulating ourselves from the power of images, implied by the very category of art.

Whereas Freedberg's work represents an art historical critique of aesthetics, moving toward cognitive science, Alfred Gell attempts a deliberate move away from aesthetic appreciation toward a sociology of objects, treating objects as social agents. Gell's attempts are similar to Freedberg's in considering the efficacy of objects as a form of practical technology: not so much what the image says or is, but what it does. Gell posits an "anthropological theory of art . . . which we can roughly define as the 'social relations in the vicinity of objects mediating social agency'" (1998, p. 7). It requires some acumen to speak of objects as agents without sounding like an "idolater." Gell does this by presenting the agency of persons as, in fact, already diffuse and "distributed" (ibid., p. 21), an effect of social networks (ibid., pp. 21–2, 36–8). He notes, in Western philosophy, the difficulties of identifying exactly what makes a person an agent, and he explains the ways in which a person is also a thing (ibid., pp. 124–5). While artifacts may be considered "secondary agents," the relation between icon and person is interactive and mutually defining (ibid., p. 21).

Ambivalent attitudes toward materiality

Some Western scholarship seems to have assumed that true religion is a matter of the spirit, not of material objects except as art or in a background role, nor a matter of the body except to transfigure it. Popularly, "true religion" involves mainly feelings and cognitive states and not interaction with objects; at the same time, excessive concern for manufactured objects represents a sign of a hollow spirituality, or of idolatry. And yet, the study of prescriptive texts does not tell us what Buddhists "believed" or "experienced," or how they behaved. Gregory Schopen has pointed out that Indian Buddhism "has been studied by modern scholars" in a "decidedly peculiar" way, because of their almost exclusive emphasis on textual and doctrinal sources (1997, p. 1). Schopen argues that the focus on doctrinal sources

> looks, in fact, uncannily like the position taken by a variety of early Protestant reformers who were attempting to define and establish the locus of "true religion" . . . Proponents of this new and historically peculiar conception of religion . . . were of necessity forced to systematically devalue and denigrate what religious people actually did and deny that it had any place in true religion. This devaluation, not surprisingly but in fact almost obsessively, focused on material objects. (ibid., 13–14)

The notion that Asians really believe that icons are alive was systematically excluded from the field of religious studies and from sympathetic or advocate depictions of Asian religions. The suggestion that Asian Buddhists "worship" the Buddha is greeted with a quick set of intricate verbal locutions meant to exclude any association of idolatry; no, they are not "worshipping" nor "asking for something," but rather "paying respect to," or even "bowing to the Buddha within."

At least some Christian missionaries were aware of the practices of installing a deity in an image, and perhaps more hazily, aware of Chinese conceptions of images. General awareness of these practices appears widely in the opening few years of the twentieth century. For example, Arthur Elwin describes the five silver "organs" found inside an image, and the process of placing this kind of animating force into images, and writes: "Every large idol in China has a soul" (1903, p. 61). This knowledge was gained, at least in some cases, by iconoclasm, as when Mary Darley poked through the debris: "When it was all gone, and nothing but splintered wood remained, I took up some little models, that had been concealed in the idol, and asked them to explain what they were, and they said, 'We thought they were his intellect and his heart.'"[17]

In many anecdotes from the mission field, the idol was not mere wood and stone, not "a lifeless thing," but had consciousness and agency. Here, the contrast was not always between the deluded heathens who thought a material object has consciousness, and the Christians who knew it as only inanimate matter. Often the contrast was between the heathens who were under the power of a cruel spirit in the icon ("Satan's seat," Boaz, 1926, p. 28), and the Christians who had gained a kind of immunity or magical protection from the spirit's power. However, missionaries often preached on the folly of idolatry by highlighting the inanimate nature of the icon's body, especially its inability to perceive or respond. Note the "mud god" motif when a missionary named Miss Codrington recounted a story of a girl frightened by "the great, ugly images and the gloom," when a woman who "no longer believed in idols" came by: "Tapping the floor of the temple with her stick, she said: 'This is mud'; then she tapped the idol and said again, 'This also is mud'" (Codrington, 1915, p. 7).

Missionaries saw icon-related phenomena and reported on it with more or less explicit condemnation or mockery; but, as we already mentioned, these phenomena were also systematically edited out of more sympathetic representations of East Asian religions. Subsequently, the phenomena were "rediscovered" by scholars, and now we have several books on Buddhist icons. There has been a kind of re-discovery of Asian Buddhist materiality, recovering what the missionaries knew all along, but without their hostile interpretations.

Sociologically, the roles of objects in Buddhism have been explained in opposing ways. According to typical textbook interpretations of the development of Japanese Buddhism, for example, the elites in the Nara and Heian periods focused their religiosity on objects because of their passion for complicated rituals and their desire to be reborn in a world of riches like the one in which they were living. We would hardly deny that some medieval elites were indeed fond of elaborate rituals and material treasures, but inevitably one sees signs of a value judgment placed on such passions—as though this ritualism and love of riches were essentially incompatible with the Buddha's Dharma, and also transgressed a ritual/religious boundary between true religion and superstition. Joseph Kitagawa, for example, wrote: "the aristocrats of the Fujiwara period were notoriously superstitious, and they enjoined every incantation and magico-religious rite" (Kitagawa, 1969, p. 80). He continued,

> Sad to say, while there were no doubt sincere clergy, many of the *Mikkyō* [i.e., esoteric Buddhism] priests were corrupt, specializing in offering *kaji-kitō* (magico-religious rites and prayers) . . . The temple treasuries fattened as these ecclesiastical organizations developed *de facto* "states within the state" with lucrative manors. (ibid.)

Kitagawa thus groups ritual, materiality, greed, and corruption, against sincerity, austerity, and purity. In contrast to the corrupt religiosity of the elites, commoners are supposed to prefer a simpler, and more authentic, piety. Kitagawa quotes Anesaki Masaharu concerning a supposedly more popular Buddhism of the Kamakura period:

> The Buddhist religion of the new age was not one of ceremonies and mysteries but a religion of simple piety or of spiritual exercise. Dogma gave way to personal experience, and ritualism and sacerdotalism to piety and intuition. (Kitagawa, 1969, pp. 109–10, quoting Anesaki, 1963, p. 168)

It is easy to recognize the modernist and more or less Protestant rhetoric that permeates sentences like this, a rhetoric emphasizing spirituality against materiality, yet it is to some extent also traditionally Buddhist. In contrast, another diffuse interpretation maintains that it is the unenlightened, ignorant commoners who are attracted to the materiality of objects, whereas the true enlightened ones (often identified with elite members of the clergy) are able to transcend it—even as they come to possess a large share of the wealth. Such characterizations are simplifications, tend to idealize a certain social class, and recognize as authentic only certain forms of Buddhism.

Whether used by an extravagant elite or a superstitious *vulgus*, sacred objects are frequently associated with an inauthentic religious experience—as opposed to a spiritual experience that does not need any material support. Religious objects are thus considered as manifestations of an unenlightened fetishism that needs to be overcome. Even a generous attempt to justify the uses of images in Buddhism and Christianity by Roger Corless reproduces the basic assumption that "spiritualities without images" are more profound than "spiritualities with images," which are prey to "misplaced reification." Images may be consistent with "higher" spirituality, but only given that "the deconstruction of images must itself be deconstructed." (Corless, 1996, p. 23)

The bifurcation of spirituality and materiality (or at least, of the essence of Buddhism and its material evidence) is not only a product of Western discursive practices; outside of the scope of Western influence, religious objects had been scrutinized as problems within Buddhism as well.

A tension pervades the history of Buddhism as both a religion of men and women who have renounced materiality on an ideal plane, and as a religion emphasizing the sacred, soteriologic role of objects on a practical level. Thus, we find statements encouraging the use of material objects (icons, relics, offerings) in worship, and at the same time criticism against such practices as degenerations that prefigured the end of Buddhism. Worship of relics is criticized in favor of worship of Dharma, but worship of Dharma in turn became a cult focused on texts as material artifacts (see Schopen, 2005, pp. 25–62; and Eubanks, 2011). Practical aspects of the interaction with objects are addressed in numerous texts, from Vinaya and related commentaries to the esoteric corpus. The tension runs through the entire history of Buddhism, and could explain the conflicted attitudes toward objects in Buddhist societies. The Chan/Zen tradition in China and Japan, in particular, is often referred to as a typical example of Buddhism's supposed iconoclastic attitudes. Well-known Zen stories present respected masters as burning Buddhist statues in temples in order to warm themselves up, claiming that those objects are not sacred, just pieces of log; or encouraging their incredulous disciples to kill the Buddha should they encounter one. These anecdotes have been interpreted in many ways, but they seem to emphasize the need for personal practice and self-realization, in contrast to formalistic devotion to sacred artifacts and reliance upon sclerotized teachings. We may interpret such manifestations of aniconism or outright iconoclastic attitudes as a remedial caution against the over-proliferation of images and monastic wealth. (See Faure, 1991, pp. 143–7; 1996, pp. 264–74; on Buddhist aniconism in general, see Swearer, 2004, pp. 24–30) Yet Chan and Zen temples in Asia have images (sometimes very many), and some are enormously wealthy; the discourse on non-attachment and rejection of objects may have been one way of dealing with these objects without actually abandoning them. Certainly the famous Buddha-burning episodes of Zen are not representative of normal life in monasteries. Chan also re-valorized the monk's body, especially the master's body, which took the ritual position of the Buddha image, and was spoken of as identical to the Buddha (at least while in that position).

Many Chinese Buddhist scholastics have reflected on the materiality of their Treasures, and on the problems of relations between their transient materiality and their timeless transcendence. For example, Daoxuan described Buddha images (of stone and wood) as the "tangible Buddha"; the paper and ink of scripture as the "tangible Dharma"; and the living monks and nuns as the "tangible Samgha" (respectively, *zhuchifo, zhuchifa, zhuchiseng*).[18] We cannot normally see the Buddha, but the Buddha-image is the "tangible Buddha." Daoxuan makes it clear that although the Three Treasures in reality transcend material form, these tangible objects are indeed the manifestations of the Buddha, and furthermore, if one is sufficiently advanced in spiritual cultivation, one can perceive the real principle in these material objects. In doctrinal terms, the difference between the ultimately true form of Buddha (his *dharmakāya*) and his specific, historical person (*nirmaṇakāya*)—and by extension the material images of the *nirmaṇakāya*—is a matter of individual perception, not reality. Sadly, he laments, the dull-minded masses see only the material aspect. In many respects, this is fairly standard Buddhist psychology: it is only our defiled perceptions, faulty cognition, and bad mental habits that prevent us from seeing Buddha everywhere.

If the assertion of the living presence of the spirit in the icon contradicted the casual observer's perception of the icon as a mere "chunk of wood," Daoxuan recalled the characteristically Buddhist notion that our experience of objects is determined by the purity or impurity of our perceptions. Daoxuan cites the encyclopedic *Dazhidulun*: "All Buddhas constantly emanate radiance and teach the Dharma; sentient beings, due to beginningless sins, meet them face to face and still do not see. Surely, if one image is like this, the other images are as a rule also thus."[19] Like many scholastic monks of his period, Daoxuan attempted to embrace all of Buddhism conceptually through a series of categorizations that asserted the non-contradiction or non-duality of the Three Treasures, and the unity of their material form and their transcendent realities. Buddha, Dharma, and Samgha are "one body," so that the transcendent and the tangible are ultimately non-dual. The doctrinal logic of non-duality asserted that all Buddhas are identical to Śākyamuni Buddha, to Buddha images, and indeed to Buddhist monks. Buddha as the cosmic embodiment of reality is, despite appearances, identical to the various more or less impressive Buddhas gathering dust throughout the land. To the eyes of the enlightened, the Buddha unchanging is the Buddha hand-crafted. This identity of true nature and expedient manifestation was a case of the more philosophical axiom that "emptiness is form." Similar arguments were made throughout East Asia; we find them developed in Japan until the early modern period (Rambelli, 2001, 2007).

And yet, despite—or perhaps because of—the transcendent claims of Buddhism, the use of religious objects was considered beneficial and was sanctioned by countless sutras and canonical texts. Perhaps, this ambivalent attitude toward objects and materiality in general originates in a central Buddhist concept, that of merit-making, and its related practices of gift-giving, envisioned as essential to securing good karma. The idea that *material* donations (objects or services) to religious institutions and their members (the Samgha) would generate *spiritual* benefits is crucial in any investigation of Buddhist materiality and the reactions against it, because material objects can be transfigured through ritual action into sacred entities. On the other hand, it is clear that the acquisition of merit (good karma) was not envisioned as providing just *spiritual* benefits. On the level of everyday life, merit was believed to result in a betterment of the *material* conditions of existence as well (what in Japan is known as "worldly benefits," *gense riyaku*). This circular transformation of the material into the spiritual and back, and of the profane into the sacred and back, lay at the basis of the sacred Buddhist economy, in which a community of renunciants survived (ideally, at least) thanks to laypeople's donations, in exchange for some sport of "spiritual" services. The intrinsically dangerous nature of this exchange was recognized by religious institutions, when they issued their warnings in the form of prophecies on the end of the Dharma. This interconnection of the profane and the sacred was also an easy target of reformers, zealots, and anti-Buddhists.

The Buddhist clergy has always been well aware of the dangerous connections between religious objects and afflictions (Sk. *kleśa*, Jp. *bonnō*). Prophecies of the end of the Dharma claim that the wrong emphasis on objects is one of the most powerful signs of the degeneration of the Samgha and of the end of Buddhism. Particularly interesting in this respect is a section of the *Da baoji jing*, in which the Buddha explains the situation during the final age of the Dharma. At that time, the text argues, the bodhisattvas will not use sacred objects (including relics and *stupas*) merely as skilful means (*upāya*) to lead the laypeople to the righteous path of Buddhism; for them, the use of objects will become a goal in itself, detached from learning, meditation, the acquisition of wisdom, and manifestations of compassion. To these degenerated bodhisattvas, worship of *stupas* and relics will be a job like any other. To them, relics will become vehicles of afflictions, suffering, ignorance, and rebirth, rather than tools for liberation.[20] Texts like this warn Buddhists of the ambiguity of sacred objects, since they can be effective tools for salvation but also terrible weapons of suffering.

However, even prophecies could be given an unexpected twist. In Japan, most Buddhist denominations, at least since the mid-eleventh century, explicitly envisioned themselves as providing ways to cope with the end of Dharma; in this context, a "deformed" and inflated use of materiality, much as described in eschatological prophecies, was regarded as the unavoidable norm of the times, rather than the exception. This might explain the importance that Japanese Buddhism attributes to material objects and to ritualized interactions with them—and also to the destructive violence that was wielded against them when religious institutions and their practices became the target of widespread anti-Buddhist sentiments.

The Fragile Dharma

Even the Dharma was not immune to a certain kind of death. Buddhists in East Asia, confronted with the fact of change, believed that as time passed, some of the Buddha's enlightening ideas, scriptures, and practices would be lost. The possibility of escaping *samsara* would dwindle because of persecution, carelessness, or misunderstanding. According to the prophetic tradition of the End of the Dharma (Ch. *mofa*, J. *mappō*), which began to circulate in the second century CE, as the echoes of the Buddha's voice gradually fade, our world will plunge into such awful degeneracy that the average life span will be only ten years. In these benighted days of corruption and abuse, the last fragments of true Dharma will be snuffed out by demonic forces. Fifteen hundred years after Buddha's Nirvana,

> the retinue of Māra and the heretics (*waidao*) will jump with joy and exultation, and will compete with each other in destroying stupas and temples, and killing monks. All copies of the Buddhist Canon will move to Kuśīnagara, and the Nāga King Anavatapta will plunge into the sea with all the sutras. This will be the end of Buddhism.[21]

Of course, this only means the end of the visible, meaningful Buddha-dharma as manifested among human beings. Such visions of the end involved the active destruction of Buddhist objects and the killing of monks—either by evil rulers or even by monks themselves:

> They [secular rulers] will steal and usurp the treasures belonging to the three jewels [of Buddhism], and they will not shrink from evil deeds. They will gradually destroy images and stupas, and the materials used for worship will diminish . . . [Later,] the monks will all kill one another, and not a single one will remain alive. . . . Mother Mahāmāyā [the mother of Śākyamuni] and the rest, together with their numerous attendants, will gather up the colored robes of the monks and carry them off. (Nattier (trans.), 1995, pp. 253–5)

The End of the Dharma is represented as a material destruction of icons, temples, and scriptures, as the confiscation of monastic robes, and as the killing of monks, resulting from the corruption of monastic practice. Monastic practice degenerates due to the loss of the Dharma's effective power and to a more general degeneration of human nature. Although the End of the Dharma was regarded as inevitable and various timetables were proposed, its precise schedule was not necessarily fixed.[22] In order to postpone the destruction of Buddhism (or, rather, of its manifestations), Buddhists urged the multiplication of temples, scriptures, and images. Many scriptures were buried, so that they could be rediscovered later. Scriptures and images were carved in stone, on cliff faces. The impending threat of the End of the Dharma was invoked in order to exhort believers to greater diligence, to promote actions that would prevent monastic corruption and violence against Buddhism, and at the same time to encourage sponsorship of the Samgha. Still, it is not surprising that a sense of inevitable decay pervades Buddhist thinking about its own materiality and indeed its own spirituality. The basic Buddhist doctrine of impermanence asserts the inevitability of change and thus the dissolution of any and all composite objects.

It is not unlikely that these prophecies about the end of Dharma were in fact reinforced by actual events. Chinese pilgrims to India such as Faxian and Xuanzang described the condition of disrepair and destruction characterizing, in their eyes, many important Buddhist sites, including Kapilavastu, Lumbinī, and Bodhgayā. (On Faxian see Legge, 1965, pp. 64, 68, 83, 87; on Xuanzang see Li (trans.), 1996, pp. 173–4, 185, 245–6) It seems that by the seventh century, Buddhism had moved away from its original sites, which were left in a condition of semi-abandonment. Furthermore, Buddhist temples in India were the objects of numerous attacks by local secular authorities and/or by mobs already long before the arrival of Muslim invaders. As Buddhist temples were centers of accumulation of wealth without autonomous military defense, they were easy targets. These attacks, which became more frequent after the end of the Gupta empire in the seventh century, were carried out by Hindu feudal lords and statelets, whose Brahmanical ideology despised the Buddhist ideals. In the eighth century, Muslim invaders began to attack Buddhist temples in what is today Afghanistan and Pakistan; Muslim armies were also responsible for the destruction of large temples in north-east India such as Nālandā, Vikramaśīla, and Odantapuri, which resulted in a mortal blow to Buddhism in the Gangetic plain.

The economic value of a golden Buddha image was a boon, because in the imagination of the worshipper its great economic capital easily carried over into symbolic capital. But the economic value of images (especially of metal) was also a problem, not only because of the perceived contradiction between the doctrinal rejection of worldly wealth and the opulence of some monasteries but also because of the ease with which Buddha images could become strings of coins, or swords. The problem extended beyond images: clerics could also be converted to tax-paying farmers or soldiers, usually accompanied by rationalizations that monks and nuns were unproductive economic parasites. Some of the impetus to sacralize the metal, paper, and flesh of Buddhism must have come from the fact that these things had economic value. At certain moments of financial crisis, the state invaded monasteries and took their metal, and disrobed the clerics (or curtailed further ordination). The *Jiu Tangshu* mentions "the assignment of the copper acquired from Buddhist images and bells to the state mint, the iron to the making of agricultural tools, and the gold, silver, and jade to the government treasury" (quoted in Reischauer (trans.), 1955, p. 225). One of the tactics employed by anti-Buddhists was to emphasize the raw materiality of images, "even going so far as to treat stone images of the Buddha as if they were bricks or tiles," as the Sui-Tang anti-Buddhist Fu Yi did (Gjertson, 1989, p. 264). Iconoclasts tend to emphasize the materiality of the objects they break, as the breaking of icons is often intended to demonstrate their mere materiality, and in a sense, the superiority of the animate over the inanimate.

Whereas very little could be done against attacks of overwhelming military power (even though Japanese temples tried to protect themselves by raising and training their own militias), Buddhists could at least try to preserve their temples in time of peace. Buddhist literature contains many injunctions aimed at protecting the material objects of the Three Treasures. These injunctions are found throughout the Vinaya (monastic codes) and in many re-formulations by Buddhist exegetes. For example, Daoxuan wrote a training manual for newly ordained monks, the *Jiaojie xinxue biqiu xinghu luyi* (Admonitions for New Student-Monks to Maintain Discipline), which is dotted with rules pertaining to the material preservation of monastic property:[23]

> Do not drive nails into the partition-walls.
> Do not write on the walls and doors.
> You must not climb over the fences or walls.
> You must always regulate the fire and candles. Do not neglect the fire.
> You must not leave a lamp or candle exposed.
> You should take good care of the material objects of the Three Treasures, and do not cause damage or loss.
> If there is damage to, or waste of, the permanent property [*chang zhu*] or to the objects pertaining to the Three Treasures, etc., you should doubly restore it.
> All who desire to handle the sutras must first wash their hands using soap.

Monastic codes and commentaries thus assumed that a certain amount of destruction, deterioration, and de-sacralization was inevitable, and included instructions on its management. Other parts of the disciplinary literature of Buddhism explicitly addressed the need to clean: to wash the feet, face, hands; to brush teeth; to sweep the floor; to wash bowls, basins, clothes, mats, towels, and socks; and to bathe Buddha images. Washing Buddha images was a form of worship, involving an altar, the use of incense, colored and scented water, and verses to be chanted. For example, the *Foshuo mohechatou jing* (the *Sutra of the Mahasattva Spoken by Buddha*),[24] begins with a description of gods washing the newborn Buddha with 12 kinds of scented water and various flowers. While re-enacting this mythic moment by pouring water on the image, you chant the verse:

> I now bathe all Tathāgatas.
> May his pure wisdom and merit inspire the assembly,
> Let the sentient beings, suffering from the five impurities, be freed from the filth,
> And witness the Tathāgata's pure Dharma body.

A similar text translated by Yijing specified that used bath water was imbued with sacred power, and must be disposed of according to rules: "Drain [the water] onto a pure place; don't let your foot step on it."[25]

The ritualization of washing points to an anxiety over keeping representations of the Buddha "spotless," free from the unintended build-up of dirt. Daoxuan's guide was concerned to prevent wear and tear or carelessness—unintentional damage. But there were also more forceful injunctions against intentional damage. In fact, according to some Buddhist texts, willful destruction of the "material objects of the Three Treasures" was one cause for damnation. In the *Kyōyōshū* ("Instructions to Foster Filiality"), a medieval Japanese text attributed to Kakuban (1095–1143), we find a list of crimes causing rebirth into the Avīci hell, the lowest of the infernal realms where the enemies of the Dharma are punished "without interruption" (Rambelli, 1999). These heinous crimes include: the five gravest actions against social order and Buddhism (*gogyaku*), that is, matricide, patricide, killing an arhat, making a Buddha's body bleed, and disrupting the Samgha; the four grave crimes committed by monks against the purity of the sangha (J. *shijū*, Skt. *catvāraḥ pārājikāḥ*) resulting in expulsion from the Samgha: lasciviousness, theft, murder, and deceit; setting fire to Buddhist statues and to the residences of monks; stealing the properties of Three Treasures; and the unlawful appropriation of sacred objects.[26] Here we see actions against the social order and Buddhism, and actions committed by monks against the purity of the Samgha, which as noted above were signs of the imminent end of the Dharma. We also see destruction of sacred images and buildings and the misappropriation of sacred objects. Material destruction of Buddhism was prominent among the gravest sins—on the same level in fact as killing monks and Buddhas. Particularly important were crimes threatening the harmony of the Samgha. Indeed, in medieval Japan destroying the Samgha was considered the gravest of the five heinous crimes because it caused disbelief and therefore prevented all beings from attaining salvation.

Buddhism is well-known for its many mechanisms for the management of violence such as its moral teachings in favor of compassion, or the various calming practices such as meditation and propitiation. In China and Japan, monasteries were built on battlegrounds to calm the ghostly violence of the war dead.[27] However, Buddhism also actively promoted discourses of violent retribution against its enemies. For example, Buddha's punishment was invoked upon the enemies of religious institutions: the enemies were presented as doomed to poverty and sickness in this life and to a fall into Avīci hell in the next (Rambelli, 2002). The idea that buddhas and bodhisattvas are not only compassionate toward sentient beings, but can also punish their enemies is to some extent extracanonical and heterodox, but was nonetheless a widespread view. However, the buddhas, being fully enlightened beings, did not punish out of spite or anger, but in accordance with the inexorable law of karma that rewards evil with suffering. Destruction was thus an important aspect of the life of Buddhist institutions, both symbolically and in concrete terms.

Buddhists felt a pervasive sense of the threat of destruction, and a concern with preservation: in a degenerative concept of time in which all things—even the teachings of Buddha—decay; in a series of meticulous rules focused on the prevention of punctures, tears, burns, and stains; in rituals fending off the accumulation of dirt on Buddha images; and in threats of damnation for attacks on the material and bodily presence of Buddhism.

However, the fundamental ambivalence toward objects prevented Buddhist institutions from developing strong discourses and effective reactions against iconoclastic occurrences. They threatened the (animated) icons' revenge against their potential attackers, and performed rituals of malediction (Rambelli, 2002). But when persecutions could not be stopped, as in the case of massive anti-Buddhist campaigns involving widespread violence, disrobing, and material losses (statues melted down to forge coins or cannons, scriptures burned to ashes, temple lands confiscated, temple buildings converted into schools or military barracks), Buddhists attempted to rationalize these events as warnings from the otherwise invisible realm of the buddhas or as punishments for the Samgha's own pollution and degeneration—in other words, acts of destructions were re-conceptualized as acts of self-sacrifice. We will address some of these themes in Chapters 2 and 5.

Ambivalence toward sacred objects' supposed animation, including considerations of the ways in which we perceive objects as "alive"—however un-mystical, materialist, agnostic, or detached such considerations may be, can be manifested also, and paradoxically, through acts of violence on objects. David Freedberg remarks: "by and large iconoclasm represents the most heightened form of making plain one's superiority over the powers of both image and prototype, of our liberation from their unearthly thrall" (Freedberg, 1985, p. 35). The mute passivity of even the most life-like icons is most obviously demonstrated by attacks and offenses against them. Similarly, violence on another person is one of the most common ways that personal agency is experienced and accentuated. Violence on objects highlights paradoxically both the passive thingness of the object—upon which the actor imposes his will, accentuating the contrast of active living agent and passive insentient thing—and simultaneously testifies to the thing as an active agent, as a presence at least powerful enough to provoke such reactions. Iconoclasm attests to the power of icons even as the iconoclast demonstrates the icon's impotency.

"Iconoclasm" in East Asia

How can we address destruction of Buddhist sacred objects and violence against Buddhists? Can they be considered cases of iconoclasm or not? Iconoclasm and its related concept of iconophobia are traditionally confined by scholars to the West, to the Christian world—or, at most, to the monotheistic religions of the Book. In fact, the study of iconoclasm has focused almost entirely on two specific moments (though "moments" is too slight a word): the Byzantine Iconoclasm and the Protestant Reformation. For example, in the first major encyclopedia of religious studies, James Hastings' *Encyclopædia of Religion and Ethics* (1913–22), the entry for "Iconoclasm" deals only with a series of cases that occurred in the Byzantine Empire.[28] More than six decades later, the equivalent field-wide *Encyclopedia of Religion* (1987), edited by Mircea Eliade, likewise limits the entry on "Iconoclasm" to the Byzantine case.[29] In both encyclopedias, the term does appear in reference to other times, places, and religious traditions, though rather incidentally. Clearly, though, the intentional destruction of sacred objects was not restricted to those two periods, or to the West. All religious traditions, especially in their encounters with other traditions and in periods of internal reform, or in haphazard, "random" ways, have experienced one form or another of iconoclasm—the breaking of icons. As long as there are things people worship, there is the (threat of) destruction of those things, and also some attempt to make sense of the victimization of the material manifestation of the supposedly imperishable. Just as religions attempt to make some part of our material world sacred, the perishability of the manifestations of divine beings has been a constant problem for religious institutions, and may help to explain why gods are so often invisible and intangible.

Furthermore, in Candea's definition of the term in the *Encyclopedia of Religion*, iconoclasm was primarily about the theology of images, a "trend of thought . . . debates" (Candea, 1987, p. 1). This use of the word limits iconoclasm to the intellectual realm, contradicting the very term itself: "breaking an icon" is primarily an action affecting an object. Thinking of iconoclasm as specific social movements also suggests that there are distinct periods with and without significant iconoclasm. For us, this is also too restricted a sense. We see iconoclasm as far more diffuse and widespread in time and place, as a permanent background (or at times, foreground) for cultural and religious activity. Although there are distinct movements of accelerated iconoclasm, "there was never a period in which images were not questioned" (Mangrum & Scavizzi, 1991, p. 5).

Other recent studies have pointed to a larger cultural context for Western instances of iconoclasm. It appears that they were almost never based exclusively on pure doctrinal and philosophical speculations about the nature of the divine and the possibility of its representation, but had significant political and economic motivations as well. Byzantine iconoclasm, for instance, was related also to other, important political and economic factors related to the status and function of the emperor as the supreme ruler invested with a quasi-divine power.[30] Thus, iconoclastic campaigns involved an economic policy targeting monasteries, which owned extensive land possessions and which were the main producers of sacred images and the beneficiaries of their veneration. Moreover, targeting the territorial control and the economic power of large monasteries was related to the ruler's centralizing policy directed against the localistic structure of the Byzantine empire, in which each locale was protected by specific saints, represented in icons and object of cults controlled by the local churches. All this had to do with the Empire's religious policy on the relations between the church and the State. Finally, foreign policy was part of the game, in particular the measures toward the growing power of Islam. (In 721, Caliph Yezid II, after his conquest of provinces from the Byzantine Empire, had all images in the sacred sites and in private homes of these territories destroyed).[31] All this demonstrates that iconoclasm was a definitive conflict in church history, as "all strata of society were involved in the struggle. The fight was violent, bitter, and desperate. The cost of victory was enormous" (Florovsky, 1974, p. 101). The eighth- and ninth-century Iconoclasm in the Eastern Church was on the one hand a series of debates over the theology of representation (and hence also relics and the status of the saints), and on the other hand a contest over the imperial control of the church. Monastic institutions were cast as the defenders of icons against the imperial government and the army.[32] Imperial and military authorities often took the lead in iconoclasm.[33] Iconoclasm was also an attempt to erase the bodies of monks (by killing them) as well as the *habitus* of monks: by banning the wearing of the monastic habit, by forced marriage, by ritual humiliation,[34] and by destroying their physical context (e.g. converting monasteries into barracks).[35]

Protestant iconoclasm also had larger political and cultural motivations beyond doctrinal issues, involving changing social attitudes and social orders based on the different conceptions of God and on representations of that new divine nature and order.[36] It would be wrong to ignore such interplay of political, economic, and social issues with religious themes also in the case of East Asia.

The category of sacred objects includes far more than icons. While icons (statues of Buddha, for example—more or less humanoid representations) are undoubtedly crucial targets, we wish to consider a larger field of destruction. The study of Reformation iconoclasm has, in fact, already broadened beyond attacks on icons. In his study of sixteenth-century Western Europe, Lee Palmer Wandel gives us a definition that names "images" as targets:

> Iconoclasm is defined by its objects, its "victims," the focus of its violence. . . . [it is] an attack, violence, either verbal or physical, usually physical, often combined with violent words, against Christian "images," predominantly, but not exclusively in churches. (1995, p. 26)

However, he then demonstrates that this violence targeted any number of church objects: "For the iconoclasts, all the objects in the churches—candlesticks, monstrances, lamps, chalices, rood screens, as well as the carved retables, panel paintings, freestanding sculpture, and crucifixes—were 'idols.'" (ibid., 190). Wandel clearly expands the meaning of "icons" to include ritual implements and more generally all sorts of objects associated with religious institutions and practices. This represents a significant move beyond received treatments of iconoclasm as a theological discussion, toward an approach to the destruction of religious objects as a more general and pervasive cultural phenomenon; in other words, this is a move from theology to culture. In recent years a number of social histories of Reformation Europe have enriched the discussion of that period of iconoclasm (among others, Duffy, 1992; Michalski, 1993; and Phillips, 1973).

Another important theoretical step has been Sergiusz Michalski's redefinition of the Protestant iconoclasm as "rites of destruction." Michalski envisions iconoclasm broadly as "hostility to religious images . . . that manifests itself in their destruction through a more or less ostentatious public act" (1993, 75ff.). Although he uses this public-performance definition to exclude "a large number of the events in Protestant lands" from the category of iconoclasm, we prefer to see the theatrical nature of acts of destruction as a common and important feature of destruction as cultural activity, rather than a definitional sine qua non.

Limitations in scope are found in art history, where the term "iconoclasm" tends to mean the destruction of *art*, and not necessarily of religious art. In a ground-breaking work, Dario Gamboni included a token chapter on non-Western destruction, while admitting that more work should be done outside Europe and apologizing for his "unsystematic," "incomplete," and "brief glance" as the subject is "enormous" and "largely unexplored" (1997, p. 107). The weakness of this one non-Western chapter in an otherwise superb book makes clear the need for better studies of non-Western iconoclasm.

A recent collective book, edited by Stacy Boldrick and Richard Clay, makes significant steps in expanding the category of iconoclasm to various historical periods, geographical areas, and cases of destruction (2007).

Thus, despite the conceptual expansion of the field of iconoclasm in recent studies, with some exceptions, most authors still consider, intentionally or not, iconoclasm to be essentially a Western, Christian phenomenon. An article by Gilbert Durand in the *Encyclopedia of Religion* does mention iconoclasm in Buddhism, primarily to define Buddhism as "iconoclastic" and "originally apophatic," despite the later "invasion" of multiple representations of buddhas and bodhisattvas (1987, p. 110).

It has been argued that East Asia has been notably free from "true" iconoclasm: there, acts of destruction against religious objects have been only random and irrational, or motivated by political and economic issues. Paul Demiéville made this argument in his seminal study (1974). Recently, John Kieschnick has reiterated this view in his study of the material culture of Buddhism in China. Kieschnick writes that the numerous campaigns aimed at destroying Buddhist images in China, from the sixth century to the Cultural Revolution a few decades ago, "were in general carried out either for the symbolic significance of the act or for economic reasons" (2003, pp. 70–1). The symbolic value of these campaigns was animated by the perpetrators' contempt for Buddhism, defined as the "teaching of icons." By carrying out systematic programs of destruction, "[n]ot only could the state deal a fierce symbolic blow to corrupt monks and superstitious peasants by smashing their most sacred icons, but the state could at the same time fill its coffers with copper, gold, and silver by melting down the bits that remained" (ibid., p. 71). In fact, Kieschnick continues,

> [a]rchaelogical evidence confirms that the primary target of Wuzong's iconoclasm was metal images; wooden statues were left alone. Similarly, in 1958, during the Great Leap Forward, monks were pressured into contributing Buddhist icons to the scrap metal drive. (ibid.)

While Kieschnick allows for non-economic reasons—contempt for a foreign religion, symbolic performance—he virtually rules out iconophobia and iconoclasm because there was no conceptual attack on iconic representation as such. He concludes:

> none of the incidents I have discussed so far reveal iconophobia in a strong sense of the term . . . The theoretical foundations of these instances of image destruction are only iconoclastic in a superficial sense, in that the perpetrators were not opposed to the notion of erecting icons in itself . . . The focus of criticism was always on what the images represented socially and economically rather than the numinous forces they contained; the state seems never to have found disturbing the Buddhist claim that icons were alive with sacred power worthy of veneration. It was a claim to be dismissed or ignored rather than challenged or rooted out. (ibid., p. 72)

In contrast, adds Kieschnick, "true iconophobia involves a preoccupation with the inadequacy of physical representation of the divine . . . More than inadequate, such idols are dangerous, for they may lead an unsuspecting devotee into false worship" (ibid.). For Kieschnick, then, Chinese iconoclasm can hardly be named such because there was no operative concept of idolatry. The definition of the term iconoclasm requires a particular doctrinal element, such as an articulated desire "to salvage a higher Buddhist truth from desecration" (p. 78).

We believe that this view does not accurately represent the complexity of motivations surrounding acts of violence against sacred objects in China and, more generally, in East Asia. First, the decision to base the concept of iconoclasm exclusively on iconophobia as something involving "a preoccupation with the inadequacy of physical representation of the divine," as Kieschnick does, is decidedly too narrow. In fact, it is possible to argue that no religious tradition, even the most iconophobic one, renounces any and all forms of representation of the sacred. As Alain Besançon has rightly indicated, iconoclasm in its traditional sense has to do with doctrines and ideas concerning the representation of the divine. The issue here is whether the divine is conceivable (whether it can be the object of representation) or not. For example, an Edo-period Japanese scholar of Western studies, Nishikawa Joken (1648–1724), discussed the early Ming critique of the religious use of images. Chinese Confucian authors argued that images cannot be exact replicas of their originals; but if they are faulty imitations, then they cannot have the power to respond to prayers. Thus, the Confucians suggested that people had to worship by using "divinity" itself (*shinshu*) instead of images (*eizō*). Joken criticized this position for two different reasons. On the one hand, he pointed to the absurdity of the claim that pre-Ming religious activities, since they employed faulty images, were always unsuccessful. On the other hand, he stressed that there exists no entity in the universe which is self-identical, as even one's self keeps changing; in this sense, there can be no distinction between divinity itself and images (1985, pp. 149–51). In this case, we see a philosophical iconoclastic critique of images countered in both philosophical and pragmatic terms. While more work needs to be done on the Confucian discourse about religious objects, it is clear that iconoclasm is in fact a continuum of different positions, ranging from those allowing only certain representations of the sacred (the cross, the Eucharist, the divine Word) to those completely denying the possibility of representation itself, such as Islam and Judaism—though even in those extreme cases popular devotion at least makes use of sacred objects (talismans, images of saints, etc.) signifying or embodying in various ways the divine. Iconoclastic attitudes also manifested different positions concerning art, and mimetic arts in particular. Calvin, for example, recognized the value of secular figurative art, but denied the value of religious subjects; Islam, especially in modern times, tends to deny the value of representation tout court.[37] Going back to the case of East Asia, authorities attacked Buddhism also in the name of different systems of representation of the sacred, variously based on Confucian, Daoist, or Shinto values, but more often than not they had an absolute vision of the center of political power and a utilitarian vision of religion as *instrumentum regni*. If this is true, we are indeed dealing here with conflicts concerning the adequacy of representation of the divine and the source of authority and political legitimacy—themes that fit well within received accounts of iconoclasm in the West.

Secondly, it is not true that intellectuals, devotees, and politicians in China (and Japan) never interrogated themselves on the nature of their icons. This interrogation was not limited only to the well-known case of a few Chan intellectuals who expressed iconoclastic ideas (which were, however, never put into practice—see Faure, 1991, pp. 143–7, and 1996, pp. 264–74), or to the Neo-Confucian aniconists who removed images from Confucian shrines in the sixteenth century (Sommer, 1994). As we have already seen, Buddhists had complex, and sometimes conflicting, ideas about the status of their sacred objects; to them, Høbjerg's idea of "inner iconoclasm" was not completely alien. (On this term, see Chapter 5.) The care taken to consecrate and maintain icons is a measure of concern for their adequacy. If an influential segment of a society can engage in systematic acts of iconoclasm despite the existence of strong religious organizations enjoying widespread popular support, that means that part of that society questioned modes, ways, forms, and goals of received representations of the sacred. If a venerated, and supposedly miraculous image can be fairly easily destroyed with impunity, that raises important questions about its sacredness and even about the nature of the entity it represents.

Thirdly, the absence of articulated philosophical concerns about the adequacy of material representations of the sacred can be taken to mean there was no "true" iconoclasm only if iconoclasm is conceived in narrowly doctrinal terms. To do so is to fall prey to what Bourdieu called the secret conviction of the thinker: "the 'thinker' betrays his secret conviction that action is fully performed only when it is understood, interpreted, expressed" (1990, p. 36). In this book we propose to take iconoclasm in a more literal sense. We will have cause to turn to the role of interpretation of destructive acts, but we do not want to arbitrarily divide acts that were economic or xenophobic or whose reasoning was never recorded, and acts that were linked to a particular philosophical rationalization. Chapter 5 is an attempt to chart out the range of possible meanings of the term.

To sum up, for us, "iconoclasm" occurs not only at certain moments or periods; not only in Europe; not only against icons (statues, pictures) but also against many other objects associated with religious institutions; not only as a matter of theology; not only as lamentable mob violence; not necessarily as something to be condemned; and not only as a problem to be solved. From our perspective, iconoclasm is not limited to doctrinal and religious discourses about icons, but can operate on several different cultural registers. An act of iconoclasm may be an act of personal purification; or a statement about religious orthodoxy and legitimacy; or a ritual to transfer objects from their physical state into the invisible world of buddhas and deities; or a direct intervention against the structure of the religious field; or a mode of interpreting "natural" or "random" damage.

Finally, our expanded definition of iconoclasm as cultural activity of destruction enables us to look at acts of destruction of sacred objects in East Asia as meaningful "texts" signifying at least some of the ways in which people in East Asia understood the sacred and its representations/embodiments and interacted with them. Still, we put the word "texts" in quotations—we do not want to define the physical act (breaking of icons) as an intellectual discourse (such as a particular doctrinal concern).

Conclusion

Let us now try to summarize the main points of this chapter. Buddhism is a religious system fully entrenched with objects. This situation should not to be treated with a modern or Protestant spiritualistic bias; we reject the absolute opposition of materiality and spirituality. As Buddhists themselves argued, "forms" (material entities and representations) are necessary in order to realize "emptiness." We also need to be conscious of an anti-materialistic bias in religious studies and Buddhist studies, related to discourses against "idolatry" and "superstition," and grounded in modern Western logocentrism (emphasis on verbal representations of the sacred—doctrines, rather than on bodily activities). In contrast, many Buddhist texts explore the issue of the status and role of objects, often sacralizing the material and economic aspects of the cult. In particular, we explored examples of Buddhist theories of material objects, namely Daoxuan on the tangible three treasures, and the materialistic outlook typical of much of East Asian Buddhism. We argued, following Baudrillard, that the numerous and various objects participating in the monastic economy reflect and serve to create a "total order" of life and the world. Indeed, materiality allows for many kinds of participation in religious practice, such as everyday devotional activities, the creation and manipulation of lineages, and even missionary work. In this respect, we also addressed the soteriological value of wealth (merit-making) and production of Buddha images as the "real" Buddhas for our world.

Yet, Buddhist attitudes to material objects are ambiguous and polyvocal. There are elite views of the masses as absorbed in their materiality and, vice versa, visions of "true religiosity" as the one practiced by the masses in their "simple piety." Objects are necessary for the practice of Buddhism, and yet we find the image of the End of Dharma (*mofa*, *mappō*) as an age in which monks are attached to materiality: objects, wealth, and prestige. On the other hand, anti-Buddhists drew special attention to Buddhist wealth as a despicable case of economic waste of resources and manpower—attacks to which Buddhists had to reply by walking on the narrow path between world renouncement and world affirmation. Materiality also made Buddhism vulnerable to attacks, and required constant maintenance of relations to the state and local elites for protection and fund raising. Esoteric Buddhism, and other schools after it, posited a sacralization of the whole universe precisely because of its materiality (the Dharma-body as the combination of the six elements and manifested in plants and rocks), but in practice attention focused on certain objects (such as relics, images, implements, and temples).

As we have seen thus far, Buddhists discussed and theorized about the materiality of their religion, in doctrines on the Buddha-body and its manifestations, including artifacts and visual representations. Sacred representations tended to be treated as manifestations in this world of entities existing in the realm of the Buddha, perhaps an attempt to save these sacred objects from their inescapable perishability. And at the same time, Buddhists also performed funeral services for their images and, especially in Japan, even developed memorial services for material objects. Thus, Buddhist thinkers devoted considerable intellectual effort in discussing and establishing the ontological status and soteriological role of sacred objects—in doctrinal arguments that are largely ignored today. Certain objects have a well-articulated special status in Buddhism: temples, for example, are often microcosmic entities, "power spots" where the Invisible becomes visible. Great care was taken to distinguish monastic space as qualitatively different from the secular world. Statues are not simple representations, and still less "idols," but animated icons, veritable living beings. Ritual implements are often mandalic instantiations of enlightenment. These objects were not merely the incidental superstructure of Buddhism, but were given explicit roles within the doctrinal and soteriological systems. Monastic space imposed specific demands of ritual, as various local interdictions suddenly lay on the body of the pilgrim at the gate. The intentional destruction or diminution of the Buddha's temples and objects was subject to particularly intense punishment by the inexorable workings of karma.

Interaction with such sacred objects was, and still is, an essential aspect of Buddhist practice and Buddhist identity. Violence was also an important mode of interacting with sacred objects as a way both to stimulate their latent power or to reduce it. "The religious power and importance of these objects are . . . only underlined by the fact that they frequently had to be forcefully removed and destroyed and always had to be fulsomely denounced with an otherwise curious ardor" (Schopen, 1997, p. 14; see also Gamboni, 1997, p. 197). Those who deny the religious value of sacred objects, or go so far as to attack them, are in fact among those who are most aware of their power. Thus, forms of iconoclasm—broadly understood—were (and still are) important ways of dealing with the sacred, and with the relations of sacred entities and matter. Materiality, and specifically religious objects, both establishes and attacks the sacred.

And yet, as Baudrillard writes, "although in the first instance objects present themselves to us as reassuring, as factors of equilibrium, albeit of a neurotic kind, they are also in the end a factor of continual disillusionment" (Baudrillard, 1996, p. 123). Despite their functions as representations of stability, immutability, and of the non-quotidian, objects are not eternal. They are fragile, they fail to function, and need constant care to protect them against dirt, decay, wear-and-tear, accidents, and all kinds of intentional and nonintentional violence. This is perhaps the major ambiguity of sacred objects: representations of an unconditioned dimension transcending the everyday but at the same time dramatically vulnerable and transient.

Part Two

Histories

Instances of Religious Destruction in East Asia

2

Iconoclasm and Religious Violence in Japan: Practices and Rationalizations

Fabio Rambelli

In April 1945, during the final months of World War II, the abbot of Heikenji in Kawasaki had a revelatory dream concerning the safety of the precinct. The next day, the main icon (*honzon*) was secretly removed from the temple and stored in a safe place. That very night, the temple was almost completely destroyed by heavy American bombings and the ensuing fires. The official interpretation at the temple is that thanks to that revelatory dream, the main object of worship, the central symbol of the temple, could be saved. Other people, however, have suggested that by taking away the icon the abbot indirectly caused the destruction of the temple.[1]

This story is a good starting point for a study of the attitudes concerning religious destruction in Japan. It reveals a tension in the nature of the temple, as being either a mere icon receptacle or an autonomous entity endowed with its own sacred power. We also see here a case of communication with the invisible realm of the buddhas (the revelatory dream) affecting the survival of the temple. Interestingly, the main icon was not able to save the temple.

Several questions come to mind. Is a temple just a container for Buddhist sacred images, or is it rather a fully-fledged sacred entity in itself? Can a temple and its icons be separated, or do they form a combined structure? If the temple is a sacred entity in itself, why does sacredness not always manifest itself as a self-protecting power? I asked these questions to a number of monks and lay Buddhists in Japan; most had never even thought of these issues, and no one could give me a satisfactory answer. Indeed, numerous scholars of religion have described in detail how a sacred place is constructed and how it operates,[2] but little attention has been devoted thus far to the *destruction* of a sacred place; a famous Japanese Buddhologist once simply told me that destruction does happen and thus it is not a theoretical problem.

However, if a major temple can be so easily destroyed, how can people trust the soteriological and protective power of its deities? What is at stake here is not a matter of naïve belief in the supernatural agency of Buddhist sacred entities, but a serious theoretical issue concerning the nature of the sacred in Buddhism and its modes of manifestation and representation in this world. Destruction seems to undermine the contract of trust binding religious institutions and their faithful. And in fact, as I will explore in this chapter, Buddhists throughout the ages have attempted to explain and rationalize instances of destruction, both accidental and intentional.

Even when a temple was burned down completely, it did not cease to preserve a certain sacredness, and in many cases temples that had been destroyed were rebuilt, sometimes even on a larger scale and with more lavish decor than the originals. In other cases, temples or parts thereof still exist perfectly preserved in museums but have lost their sacred nature. These phenomena raise several questions concerning the cultural history, semiotics, and ideology of Buddhist representations of the sacred in Japan: the doctrinal and philosophical status of sacred objects; the typology of and reasons for destruction; and the ideological role of destruction in narratives about Japanese cultural tradition. Why is a sacred place so often reconstructed after destruction? Is it enough to destroy the building in order to annihilate a temple? Or does this operation rather require additional labor, one that targets the very signification of the sacred place? What were the ideological uses of destruction?

In this chapter, I will focus on a number of cases of malevolent destruction of religious artifacts within the Japanese Buddhist context. Tackling these issues will enable us to begin to assess the place of destruction, and violence in general, in representations of Japanese cultural tradition, which is usually based instead on notions of "harmony" and unbroken "continuity." I will present some of the ways in which Buddhist religious institutions used their doctrines to explain dramatic practical events such as the destruction of temples.

We do not usually think of the destruction of a temple as an act of iconoclasm, perhaps because temples are commonly understood not as representations of the sacred, but as mere "containers" of images and sacred objects. This misperception has at least two reasons: linguistic and historical. On the one hand, language represents metaphorically a temple as a "container," as a building in which an "object of worship" (*honzon*, that is, an image) is placed—or, literally, "reposes" (*anchi*). However, this metaphor of the temple as a container should not blind us to the evidence, provided by elite and popular texts alike, that temples were "iconic" in the sense that they represented a certain notion of order and of the sacred. In spite of that, iconoclasm has been historically discussed in the West mostly as connected with images, not with architectural forms; this approach has been transposed to the East Asian context. The theoretical status of the temple, as a specific kind of Buddhist sacred artifact, was also defined within the framework of the Buddhist attitudes toward objects outlined in Chapter 1. Temples also had a complex doctrinal status, oscillating between places of worship (a worship that is in principle independent of its place and its performer), and sacred sites in which buddhas, bodhisattvas, and other entities in the Buddhist pantheon manifested their holy presence. Burning a temple, especially an important one, that claimed to exist in order to protect the state and the social order, was a challenge to the entire Buddhist system of construction and representation of the sacred. The issue was, how can a temple protect the state from the menaces of the invisible world if it can be so easily destroyed?

Domesticating destruction

The dramatic history of destruction that affected most Japanese temples is part of received knowledge, but is rarely discussed and studied. Reconstruction works altered, often in remarkable ways, the structure, shape, and size of the temples. The vast majority of temples in Nara and Kyoto (including famous sites such as Kiyomizudera, Kōfukuji, and Tōdaiji) are mainly composed of what is left, after the late nineteenth-century anti-Buddhist persecution, of edifices rebuilt during the Edo period (mostly between the seventeenth and the eighteenth centuries) and preserve a distinct Edo architectural style. Other temples were rebuilt even more recently, after World War II, and often in a contemporary reinterpretation of traditional styles, such as the Heikenji in Kawasaki, commonly known as Kawasaki Daishi. In this respect, the Kinkakuji seems to be an exception: rebuilt in 1950 after a vandalistic act of arson, it is a faithful replica of its previous shape. Thus, religious institutions had to cope with destruction by developing rules to prevent it, technologies to stop it (or at least to salvage as much as possible from targeted temples), and rationalizations to explain it in Buddhist doctrinal terms that were not self-defeating. They also developed efficient propaganda strategies that enabled monastic groups to raise funds to repair and rebuild damaged temples.

How do temples today present their complex history marked by violence and destruction? I have decided to look for a first set of clues not in scholarly works (in which, by the way, sustained study of destruction is remarkably absent), but in a sampling of brochures and other materials given to tourists visiting famous temples in several places in Japan. In general, there is a frequent use of euphemisms or partial truths. Some temples do not even mention past acts of destruction that affected them. The guide–brochure of Kinkakuji, for example, only says that "During recent years . . . the structure has become damaged, and has required extensive repairs [. . .] By October 1987 [it] had been restored to its original splendor." As I just mentioned, the temple was completely burned down in 1950. Hasedera in Kamakura also does not mention any past destruction, but presents its origin narrative going back to "June 18 of 736." Other materials mention destruction by fire, such as the Sanmon gate of Nanzenji (burned down in 1447 and rebuilt in 1626). The Tōdaiji is quite frank, stating that it "twice became victim of war, in 1180 and again in 1567. On both occasions many of the buildings, not excluding the Daibutsu-den, the majestic hall enshrining the great Buddha Vairocana, were burnt to the ground." However, the Tōdaiji also emphasizes that past acts of destruction have not affected its grandeur and glory. The entrance ticket in English says,

> The Daibutsu-den that can be seen today was reconstructed in 1708, on a scale that is about two-thirds of its original size; nevertheless, it is still the largest wooden building in the world. The Great Statue of the Buddha too has had to undergo various repairs from time to time, but it remains the largest bronze statue in the world and . . . is widely admired and worshipped.

Ryōanji in Kyoto, world famous for its rock garden, has a tourist brochure written in both Japanese and English. Whereas the Japanese text reports two instances of destruction, one caused by war during the Ōnin Disturbance (mid-fifteenth century) and the other by fire in 1797, the English version is completely silent about destruction and emphasizes instead that the rock garden in the temple "is acknowledged to be one of the masterpieces of Japanese culture." The brochure of Ninnaji in Kyoto emphasizes the temple's historical heritage and its past connections to the imperial family. The English page says: "Ninnaji: Old Imperial Palace—The World Heritage," tracing a continuity between the Japanese past (in which the temple was an imperial *monzeki*, an institution headed by a monk who was a son of the ruling emperor) and the international present (in 1994 Ninnaji was designated as a UNESCO World Cultural Heritage Site). There is no reference to past destruction, only an Edo period reconstruction is mentioned. The entrance ticket, however, mentions two reconstructions: one during the early Edo period, and the other between the late Meiji and early Taishō eras (1910s) following a fire that broke out in 1887. The English booklet of Daishōin in Miyajima laconically says that "The present building [Hondō, 'Main Hall'] was rebuilt in 1955." Kawasaki Daishi is one of the few temples that do mention wartime destruction; its bilingual guide booklet says that the Great Main Hall (Dai Hondō) "was destroyed by fire during the war in April 1945. It was rebuilt in 1958." The temple is obviously very proud of its reconstruction: "The hall . . . is a splendid cathedral of the Showa period with an architectural style of the Heian period with a modern sense." A Heian-period style "with a modern sense," a Buddhist temple called a "cathedral"—these are important aesthetic and religious issues that warrant further analysis.

Other temples choose to downplay the effects of destruction. The Kōryūji claims that its Lecture Hall (Kōdō), completely rebuilt in 1165, is the oldest building in Kyoto today. However, on both sides of the façade there are windows in typical Zen style popular since the Muromachi period—an indication that the external appearance of the building was modified well after its reconstruction in the twelfth century. In this case, it is not clear what it means for a building to be the oldest one in town. The Hōkōji, the former site of the Kyoto Great Buddha (Daibutsu)—the largest Buddha image ever made until then in Japan—originally built by Toyotomi Hideyoshi in the late sixteenth century as part of his outlandish plan for self-deification, was destroyed several times, the last and definitive one in 1973 (on the history of this temple, see Murayama, 2003). A small Buddha image, originally placed inside the head of the Daibutsu in correspondence of the center of Buddha's forehead (*miken*) survived total destruction in 1973, and is now on display at what remains of the temple as a sort of memento of the Great Buddha and at the same time, to signify that it was not completely destroyed, since the small icon represents the enlightenment of the large, extinct one.

Even from this very unsystematic analysis it is clear that temples tend to ignore or downplay acts of destruction, or in any case they treat them euphemistically; the causes for destruction are almost never presented in a clear and straightforward way. On the other hand, the temples do emphasize their old age, successful rebuilding, and value as cultural heritage sites. Temples present visitors a picture of historical success: in spite of destruction, they have been rebuilt, have prospered, and are now important places that represent the Japanese cultural tradition. When they do refer to acts of destruction, they only mention the most dramatic events in Japanese history textbooks, such as the late twelfth-century war between the Taira and the Minamoto families, the Ōnin Disturbance in mid-fifteenth century, and Oda Nobunaga's conquests—thus strengthening their self-representations as special places in Japanese history, monuments of a heroic past. Destruction as an everyday risk, which was one of the motives at the basis of the premodern Buddhist discourse of self-legitimization, is now almost completely erased from explicit public discourse. In fact, the place to look for a temple's discourse on the danger of destruction is not necessarily in its narratives but in the many no-smoking signs posted around its compound and in the prominent placement of fire extinguishers.

The practice of iconoclasm in Japan: General trends

In order to overcome such romanticization of history, which forgets the actual historical past, we have to take a closer look at concrete acts of destruction that affected temples and sacred objects in Japan. As is easy to imagine, temples and other artifacts were constantly under threat by wear-and-tear, neglect, fire, warfare, and natural disasters; since ordinary maintenance required enormous resources and labor, it was not unusual that temples could not afford to preserve themselves, so they fell in disrepair and were eventually abandoned, even for long periods. In fact, most major temples in Japan (including those on Mount Kōya and Mount Hiei, Tōji, and Hasedera) have gone through a phase of a more or less total abandonment, which lasted for decades or even for centuries. According to the logic of self-legitimization employed by Buddhist institutions in Japan, if taken literally, the abandonment of a major temple, founded by a patriarch with the official support of the emperor, was a dangerous act potentially threatening the survival of Buddhism and thus the order of the State and society. It is also for this reason that, beginning from the mid-Heian period, many ancient temples lying abandoned were restored and revived as part of large-scale activities to counter what was believed to be the beginning of the final age of Buddhism (Hayami, 1991); these activities continued throughout the Kamakura period.

I will return to the Buddhist theoretical rationalizations of destruction. Here I would like to focus on the most significant cases of iconoclasm in Japanese history. Since these acts were quite common, it would be impossible to discuss each and every one of them in detail. Instead, I present what can be envisioned as general trends and forms characterizing iconoclasm in Japan.

Throughout the history of Japanese Buddhism it is possible to identify three different modes of malevolent destruction, which I will call, respectively, medieval, early modern, and modern. These are to be understood as positions on a continuum of iconoclastic attitudes, with mutual similarities and overlappings, rather than three completely distinct ways of breaking sacred objects. It goes without saying that destruction per se as a merely physical, bodily act, despite its variations, remains essentially the same in these three modes; the actions of dismantling or burning a temple in themselves are not different. What differs, though, are the goals of destruction, the scale of destruction, and the situations in which intentional destruction was authorized—all factors that are, to a large extent, culturally determined. The first modality of destruction runs through the ancient and medieval eras; acts of destruction targeted specific sites, people, or groups within temples for specific reasons (related to doctrinal reasons, but also to economic and political motivations); there was never any attempt to attack entire sects or even the totality of Buddhist religious institutions. The situation was to change drastically toward the end of the middle ages and the beginning of the early modern era, when political entities on a regional scale (headed by daimyo, regional feudal lords) first, and the state later, began to carry out large-scale attacks on Buddhist institutions. Initially, these were limited to certain sects and temple complexes, but subsequently the entire Buddhist field became a target of iconoclastic policies. The most spectacular and cruel acts of this second iconoclastic attitude were the destruction of Mount Hiei and Ishiyama Honganji by Oda Nobunaga and of Negoroji by Toyotomi Hideyoshi, and the destruction of Buddhist temples in entire regions by so-called Christian daimyos during the last quarter of the sixteenth century. Along this line, but on a much larger scale, the anti-Buddhist persecutions of the early Meiji era, which opened up Japan's modernization, eradicated forms of traditional religiosity dating back to several centuries. This is also the first occurrence of the third form of iconoclasm. One aspect of it is perhaps more subtle and typical of the modern era; in it, religious entities are denied their traditional sacredness and re-signified as "art objects," valuable artifacts of a different order.

Medieval iconoclasm

I define medieval iconoclasm as the acts of damage and destruction of sacred artifacts that took place approximately between the mid-Heian period (second half of the tenth century) and the mid-sixteenth century. Before that, except from the first period following the introduction of Buddhism to Japan, very few significant acts of destruction have been recorded. The main feature of medieval iconoclasm is the limited and case-specific nature of destructive acts—usually, the result of localized struggles within the same temple and sect, or between different temples and sects, or between religious institutions and representatives of secular power; there was never the attempt to completely obliterate the existence of a temple or a sect, or to wipe out the Buddhist presence from a region—not to mention the entire country. One of the most notable features of the history of Japanese Buddhism between the eleventh and the sixteenth centuries is the constant state of conflict within, and among, religious institutions. Conflicts involved various groups within the same temple, temples belonging to the same sect, and temples belonging to different sects.

Infra-Buddhist iconoclasm

Major temples were staffed by a large number of clergy; increases in population and specialization, together with class and other social distinctions, resulted in a series of struggles and conflicts. Generally speaking, the clergy (*daishu*) was divided into three major groups: scholar monks (*gakuryo* or *gakushō*), often from aristocratic and wealthy families; ritual specialists (*gyōnin*), people of low- and middle social background; and laborers (*dōshu* or *shuto*). On Mount Kōya there were, in addition, a large number of itinerant religious figures (the so-called *Kōya hijiri*). A few large religious institutions gradually developed forces of self-defense—militias composed in certain cases of thousands of people. These militias, known as *sōhei* (literally, "monk soldiers") in modern times, were most of the time not permanent and professional, but tended to include all members of the clergy willing and able to participate in fights to protect the prestige and the interests of their own temples or allied institutions (see Adolphson, 2007). Especially famous were the soldier monks of Mount Hiei, who exercised a military control over Kyoto and the surrounding areas for several centuries. Large monastic institutions controlled two different kinds of militias: one based in Buddhist temples (often called *shuto*, temple laborers, but in fact composed of priests from all hierarchical distinctions, from aristocrats to outcasts), the other based in Shinto shrines (usually known as *jinin*, "men of the *kami*"). Such armed forces organized by Buddhist institutions were deployed not just to preserve public order inside large monastic complexes; they also carried out military actions against other Buddhist institutions, the state, and unruly peasants. These military practices are peculiar to Japanese Buddhism. Ethical and historical considerations aside, all this indicates that forms of iconoclasm (practices aimed at damaging and destroying religious artifacts in the name of alternative visions of sacredness, in turn based on ideas of orthodoxy, legitimacy, and authority) were deeply embedded in premodern Japanese Buddhist culture.

Especially on Mount Hiei, but also on Mount Kōya and at the Kōfukuji in Nara, scholar monks, usually a minority of the clergy but in control of the temple administration, were opposed to the lower-class majority. Conflicts among them, whether of a political, social, or administrative nature, often escalated into violent struggles involving several thousand people and ending up in acts of destruction. Usually, residences (of the abbot, the scholar monks, and the laborers) were targeted and destroyed, but at times violence spread to the entire temple compound with disastrous results. Particularly serious was the confrontation that opposed the *gakushō* to the *dōshu* on Mount Hiei between 1178 and 1179, which ended with a devastating defeat of the former party (Adolphson, 2000, pp. 159–61). At Kōfukuji, the internal conflicts between lower ranks of religious servants and the scholar monks in the twelfth and thirteenth centuries served perhaps as a prelude for larger social changes that occurred in the fourteenth century, when local estate managers and warriors successfully challenged the central elites (ibid., p. 226).

Among the struggles between temples of the same sect, the most important one was certainly the rivalry opposing the two most powerful temples of the Tendai sect, Enryakuji and Onjōji (Miidera), especially from 993 until the mid-fourteenth century; Onjōji, militarily and economically weaker, was completely destroyed several times. On Mount Kōya, the clergy of the leading temple Kongōbuji clashed repeatedly with the followers of reformer monk Kakuban (1095–1143) at their newly arising monastic center Daidenpōin between the first half of the twelfth century and the mid-thirteenth century, when members of the latter party moved to another temple, Negoroji; there they established a reformed branch (Shingi) of the Shingon sect. Before that, the Daidenpōin suffered repeatedly from acts of damage and destruction.

Particularly long and virulent was also the conflict opposing the Tendai and the Hossō sects. The conflict had a doctrinal basis, and began when the Hossō monk Tokuitsu severely criticized the Tendai sect, newly established by Saichō in the early ninth century. Doctrinal debates ensued; particularly well-known is the imperially sponsored Ōwa debate of 963 (see McMullin, 1989; Groner, 2002). Violent conflicts erupted between Enryakuji and Kōfukuji for the control of two of their branch temples, respectively, the Kiyomizudera (located in Kyoto and therefore in the Enryakuji sphere of influence but belonging to Kōfukuji), and Tōnomine (a temple situated in the Yamato Province controlled by Kōfukuji but belonging to Enryakuji). Particularly serious were the events in 1112, provoked by the appointment of a Tendai monk as the abbot of Kiyomizudera; in 1173 Tōnomine was burned down in a dispute over an estate.

Since the early twelfth century, Tendai soldier monks were also involved in the destruction of temples, priest residences, and halls belonging to new movements such as the Nenbutsu (Pure Land), Zen, and Hokke (Nichiren's followers) denominations. Members of new, reformed movements, in particular radical Amidists, sometime attacked and desecrated temples and shrines of the more orthodox sects (see Rambelli, 2004). The medieval patriarch of the Ikkō sect, Rennyo (1415–99), also displayed iconoclastic tendencies. He was strongly critical of sacred images from outside his own Pure Land tradition; he even burned "the objects of worship and other articles not in accordance with the tradition [to heat the water] whenever he took a bath" (Yasutomi, 2006, p. 24). He privileged painted images to sculpture, and calligraphic renditions of the formula "Namu Amida Butsu," known as *myōgō*, above all other representations of the sacred (Yamamoto (trans.), 1968, pp. 29–30).[3] This is an interesting case of iconoclasm based on doctrinal considerations stressing the limits of a figurative representation of the sacred and gesturing toward visual abstraction on the one hand and unmediated prayer on the other. Rennyo's order to have heterodoxical icons burned was among the causes of an attack on his Higashiyama Honganji in Kyoto by Enryakuji monks in 1465 (Daiki, 1995).

Interestingly, during the middle ages there were relatively few conflicts between religious and secular institutions. The most significant was certainly the one opposing the Taira clan against the Tōdaiji and Kōfukuji in Nara, which ended with the complete destruction of the two temples in 1180. This is one of the most dramatic pages of Japanese history, and it was masterfully recounted in the *Tale of the Heike* (*Heike monogatari*) and *Genpei seisuiki*. Sources report, however, that the Taira army did not mean to destroy the sacred sites, but simply to punish rebellious monks; the fire that destroyed a large part of Nara was probably a collateral damage, an unintentional result of what was supposed to be a much more limited military operation. Later, local officers of the Kamakura bakufu were often in conflict with religious institutions over land control, but such conflicts never escalated into an open and generalized struggle. Contrasts also arose between the imperial court and religious institutions, often resulting in violent confrontations in Nara and Kyoto; but the parties involved always tried to quickly reach compromise and pacification. I will address these events more in detail below.

Popular rebellions (ikki)

Popular rebellions (*ikki*) at times also targeted temples and what they represented.[4] For instance, some revolts actively involved personnel from religious institutions (*dōshu* and/or *jinin*), others in contrast targeted temples and shrines. Particularly extensive were the revolts, known as *Ikkō ikki* (see Kanda, 1991; Tsang, 2007), led by people affiliated with the Ikkō sect (present-day Jōdo Shinshū), and those fomented by groups inspired by Nichiren (*Hokke ikki*), which involved thousands of people in entire regions for many years. The Ikkō sect controlled the region of Kaga on the Sea of Japan (present-day Kanazawa Prefecture) from 1488 for almost a century. Its headquarters, the Ishiyama Honganji, was also a major player in military and political events during the second half of the sixteenth century, until its destruction by Oda Nobunaga. In the fifteenth century the Hokke sect spread in Kyoto and began to organize citizens there for self-defense and political self-determination, but their activities were strongly opposed by Enryakuji (which claimed control over the capital), the Ashikaga shoguns, and several regional daimyo. In 1532 mobs affiliated with the Hokke sect in Kyoto attacked the Yamashina Honganji of the Ikkō sect; this struggle escalated into a larger disturbance (Tenmon Hokke no ran, "Hokke disturbance of the Tenmon era") that ended with a devastating defeat of the Hokke organization in 1536, from which it never recovered. Peasants fought against exploitation and excessive taxation by the temple owning their lands, but also for other reasons such as traditional rights and social prestige. From the second half of the fifteenth century until the mid-sixteenth century, numerous peasants' revolts (*do ikki* or *tsuchi ikki*) occurred; in 1441, 1462, and 1490, among others, protesters occupied major temples and shrines in Kyoto (such as Kitano Tenmangū, Gion, and Tōji) to request the establishment of a virtuous government (*tokusei*) (Aoki et al. (eds), 1981, pp. 254–5); in 1486, during a struggle to end one of these occupations, the Tōji was badly damaged and a full reconstruction took more than one century.

Self-destruction and self-damage

A peculiarly medieval way of destroying temples and religious artifacts—or, at least, to question their sacredness—was self-inflicted. During a particularly serious conflict, members of one of the parties involved would occupy a temple or a hall and barricade themselves inside (*heirō*). They would threaten to set the building on fire if their reasons were not accepted—which happened in several cases.

One of the earliest examples of self-destruction occurred in 1260, when monks on Mount Hiei polluted and destroyed their own shrines in protest against the court's reluctance to prohibit the establishment of an ordination platform at rival Onjōji. Four years later, for the same reason, they set several of their own buildings on fire, including important structures such as the lecture hall and the Tendai ordination platform (Adolfson, 2000, p. 270). In 1269 monks damaged some of the sacred palanquins, receptacles of the *kami*, in front of the main building of the mountain, the Konpon Chūdō. Between 1308 and 1313 there were other protests against Emperor Go-Nijō (r. 1301–8)'s decision to bestow a highly honorific posthumous name to Yakushin (827–906), an important abbot of the Tōji and the Tōdaiji (as a result, the emperor had to relent). In 1308, for example, the Enryakuji clergy initially barricaded themselves inside the Konpon Chūdō, but then they left it and burned the Hie shrine.

This kind of protest might have unpredictable consequences. In 1297 the Tendai priest Enne barricaded himself with a group of fellow monks in the Hie Hachiōji shrine to protest against the corruption of the sect's leadership. A few months later, at the beginning of 1298, Shōsan, an associated of the Tendai abbot (*zasu*) Sonkyō, invited the rebels to stop their protest in exchange for freedom, but did not keep his promise; instead, he turned Enne in to the shogun's police and killed all his followers without mercy. An uprising ensued, but Enne's supporters lost again. Shōsan's party, in order to blame the enemies, set on fire some sacred buildings in the central compound of Enryakuji—with the devastating result that the entire temple compound was burnt to ashes (Kuroda, 1980, pp. 183–4). As Mikael Adolphson explains,

> The purpose of such self-destruction was twofold. First, it symbolized the damage that, in the eyes of the clergy, the court had inflicted on Enryakuji and, by extension, on the state itself. Second, by damaging their own shrines and palanquins, the clergy disturbed the *kami*, further adding to the distress of the court nobility. [. . .] Even when the court obliged the protesters, further measures were often deemed necessary to placate the gods, in particular, pilgrimages to the shrines of the *kami*. (Adolphson, 2000, p. 270)

According to the doctrine of the mutual dependence of the imperial law and Buddhism (*ōbō buppō sōi*), prevalent at the time among the court and in the capital region, the temples were responsible for protecting the state and maintaining its socio-cosmic order in exchange for material support from the government. Given such interdependence of state and religious institutions, the destruction of Enryakuji would also result in the decline of the court itself. As Adolphson writes, "by destroying their own property, the monks used a forceful form of symbolism to show the damage the policies of the secular elites caused the temple, Buddhism, and the state itself" (Adolphson, 2000, p. 273). This idea was further connected to dominant doctrines on the end of Buddhism (*mappō*), which required strengthened devotional activity to at least slow down the degeneration of the social order.

Another form of self-inflicted damage, which was symbolically, if not materially, almost as powerful as self-destruction, was the closure of a temple or selected cloisters by the clergy, who thereby ceased to perform their spiritual duties for the state and its leaders. Closure of a temple would also signify, in fact, the decline of Buddhism and therefore that of the state as well. It seems that closure as a strategy of self-assertion was preferred by Onjōji, rather than by Enryakuji, which would rather stage elaborated and ritualized protests, known as *gōso*, in Kyoto (Keirstead, 1991; Adolphson, 2000, pp. 240–87). Onjōji performed important closures in 1163, 1242, 1257, and 1260 when the Bakufu sent warriors to protect the temple because it feared that the Enryakuji clergy might burn the unprotected compound (ibid., p. 281). Closures were also enacted on Mount Kōya; the first instance is reported in 1162, when the clergy left Kongōbuji in protest against the imperial court's unfavorable attitude toward the temple in a dispute over land. Later, during the process that led to the temple's acquisition of Ategawa estate, the clergy even threatened in 1278 to abandon the sacred mountain (ibid.). Tōdaiji and Tōji also enforced closures, most notably between 1308 and 1312 to protest the court's decision not to honor the previously mentioned Shingon monk Yakushin following Tendai's protest (Abé, 1999, p. 373). As Adolphson reports, closure

> is still in use among several temples in Japan. In 1986, when the government threatened to impose a new tax on temples, several major tourist attractions such as Kinkakuji and Ginkakuji closed their gates to the public in Kyoto. In another more selective protest in 1993, Kiyomizudera prohibited visitors from a certain hotel to enter its premises because the hotel's plan for a new annex would ruin the scenic view from the temple. (2000, p. 404 n. 101)

It is important to note that self-inflicted damage was one of the strategies of self-assertion and self-defense of the same order as forceful demonstrations (*gōso*), in which important temples mobilized powerful sacred symbolism to display their *raison d'être* and protected their interests. However, only Enryakuji and Kōfukuji staged forceful demonstrations in the capital, whereas self-destruction seems to have been a strategy unique to Onjōji (see also ibid., p. 283). These forms of self-inflicted damage were essentially analogous to cases of humiliation of saints or relics in medieval Western church studied by Patrick Geary. Through the threat, or the actual performance, of self-destruction the temples wanted to "show how . . . the enemies had overturned the divine order" (1994, p. 148) and had therefore questioned radically the role of the temple (and, in some cases, of Buddhism in general). Religious authorities aimed essentially to put "pressures, both on the individuals in conflict and on the wider community . . . to bring about negotiations" (ibid., p. 150)—and, at the same time, to reassert the essential role of the temple itself and the justness of its requests. From another point of view, self-inflicted damage was also a last-chance kind of self-defense, which implicitly acknowledged the temple's powerlessness. For this reason, the clergy never took explicitly self-damaging initiatives if they were not certain of the positive outcome. It is significant that secular authorities almost always consented to the clergy's requests when such strategies were put in place; they even rebuilt structures and sacred objects that were self-destroyed.[5]

We should also emphasize that self-inflicted damage and self-negation, while parts of ongoing struggles concerning secular issues, also mobilized symbols and images directly related to the nature of the sacred (embodied by religious institutions) and its representations. The affective power of self-destruction, and its efficacy, was precisely its intrinsic iconoclastic nature—its utter negation (at least potentially) of the meaning of religious institutions and their symbolic system. As we have already seen, the dominant idea in Japanese politics until the late middle ages was that of the mutual dependence of religious and secular institutions and their respective orderings of reality (*ōbō buppō sōi*); religious institutions, however, claimed a symbolic and moral primacy (Kuroda, 1996). Any threat—even just perceived threats, not necessarily real ones—against the prestige, social standing, economic power, and influence at court and in the countryside of religious institutions was a threat to their primacy that might have resulted in jeopardizing the entire balance between the secular and the sacred. Seen from the logic of religious institutions, everything that was owned or claimed by them was sacred, and all acts of violence against them were potentially iconoclastic, as they threatened the status, role, and significance of the sacred institutions, their objects, and the clergy.

An example of the logic governing conflicts between the religious and the secular spheres in medieval times is the following occurrence as told by the *Tale of the Heike*. After losing a battle in 1177, in which sacred images had been hit by arrows and several monks had died, the clergy held a general assembly at Enryakuji. They decided unanimously to burn every building on Mount Hiei and abandon the sacred mountain. The court was worried about the possible outcome and sent an envoy to Enryakuji to attempt a mediation. The envoy said: "The lawless violence of the monks was inspired by a devil; the imperial sanction is indeed none other than that of a wise ruler who leads you to the land of enlightened rebirth" (Kitagawa & Tsuchida (trans.), 1975, vol. 1, p. 72). The monks agreed and desisted from their plans. The envoy was cleverly trying to excuse what the court saw as unethical behavior of the clergy; but his words signify that wrongful behavior, even by the clergy, was seen as being inspired by devils. On the other hand, violent reaction against the monks' unruly behavior was the duty of a "wise ruler" and a highly meritorious action. Here, we find an interesting logic of reversal. The clergy saw in their lost battle a form of self-defense of Buddhism (a "sacred" action), but the Court understood it as the work of evil spirits; at the same time, military aggression against a religious institution, in which monks had been killed and sacred images damaged, was envisioned as a display of the ruler's wisdom and as a form of salvation. This reversal shows that the sacred and the secular were closely interrelated, and sometimes it was difficult to decide which was which. In such a situation, conflicts were fueled and rationalizations were continuously necessary.

Considerations

In all these cases of violence and destruction against religious institutions in premodern Japan described so far, we can identify the three aspects of the religious riot as defined by Natalie Zemon Davis, namely, the "defense of true doctrine and the refutation of false doctrine," "purification aimed at ridding the community of dreaded pollution," and a certain political goal, as the destroyers believed that they were "acting in place of the government" (1975, pp. 156, 157, 161). Infra-sectarian and infra-Buddhist conflicts were almost always justified as aimed at protecting a true and correct form of Buddhism against heretics, impostors, and corrupt people. It was necessary to destroy any form of "fake" Buddhism because its spread would speed up the End of Dharma and the ensuing collapse of the socio-cosmic order. Enemies were thus envisioned as polluted/polluting, dangerous factors of communal degeneration; at times, they were even described as emissaries of Māra, the arch-enemy of Buddhism. Finally, the function of the government was to protect Buddhism (the "true" one) as the only way to secure an ordered and ethical society. If the state did not act, religious institutions (but also individuals and groups) felt empowered to attack their enemies to redress the situation. It is in this light that we can interpret warriors' interventions against religious institutions, but also the acts of rioting groups and, of course, interventions of temple militias against other temples and, more rarely, secular institutions. In other words, premodern iconoclasm targeted "fake" icons—or, more generally, representations of the sacred in any form that had become (or were threatening to become) degenerated or useless. A good example of such attitude is reported in the *Tale of the Heike*. In 1165, warrior-monks from Enryakuji assaulted Kiyomizudera and razed all the buildings. The morning after the temple had been burned down, a notice appeared in front of the main gate stating: "Ha! What is the matter with Kiyomizu Kannon, who once proclaimed that she could transform a pit of fire into a cool pond?" (Kitagawa & Tsuchida (trans.), 1975, vol. 1, p. 41). This is an ironic reference to the supernatural powers of the main icon of the temple, the bodhisattva Kannon, as described in the *Lotus Sutra*—powers that did not materialize to save the temple from the flames.[6]

Another instance of the Buddhist interpretation of the role of religious institutions vis-à-vis the state is Onjōji's attempt to form an alliance with Kōfukuji against the Taira regime—an alliance that ended up causing the destruction of both temples. The *Tale of the Heike* reports that the monks at Onjōji wrote the following message to their counterparts at Kōfukuji:

> The Buddha's Law is supreme: it protects the Imperial Law. Indeed, the continued eminence of the throne depends on the Buddha's Law [. . .] [Because of Taira no Kiyomori's usurpation,] both the Buddha's Law and the Imperial Law are on the verge of being destroyed. [. . .] We pray you to give us your aid. We will put an end to Kiyomori's evil ways, and thus preserve the Law of the Buddha. (ibid., p. 255)

The Kōfukuji clergy responded:

> Though we have established different doctrines—. . . the Tendai and the Hossō—. . . these doctrines . . . derive from the one pure teaching of the one Buddha. [We] are equally disciples of the same Buddha. Therefore, we must join together to destroy the enemy of the Buddha, though he be as mighty as Devadatta. (ibid.)

Here, Onjōji claims the preeminence of monastic institutions against the state. Kōfukuji agrees and adds that the state's attempts at reorganizing the power relations between state and Buddhist institutions, and also within Buddhist institutions, were deeds as wicked as those of Devadatta, one of the early enemies of the Buddhist Samgha.

Early modern iconoclasm

Early modern iconoclasm refers to instances of destruction of sacred artifacts that took place between the end of the sixteenth century and the anti-Buddhist persecution of the early Meiji era (1868–71). While at this time infra-Buddhist struggles and localized conflicts between secular and religious institutions decreased in importance, secular military authorities carried out campaigns against either specific Buddhist denominations or even Buddhist institutions in general, first on a regional basis and subsequently on a national scale. Oda Nobunaga's complete destruction of the Tendai institutional center on Mount

Hiei and his defeat of the Ikkō sect, for example, opened the way to important changes in the doctrinal situation of Buddhism and its territorial distribution. The attempt by a few Christian daimyos to wipe out Buddhism from their lands was something unprecedented in Japanese history; while it was used to justify anti-Christian policies already under Toyotomi Hideyoshi, it became a precedent for subsequent anti-Buddhist policies implemented locally during the Tokugawa period. Many of these policies resulted in loss of lives and destruction of precious artifacts and valuable documents.

Secular institutions against Buddhism

The early modern era in Japan is characterized by two different but complementary trends in the realm of religion. The first is the sustained attempt by a number of secular authorities, thinkers, and activists, to limit and repress the independence of religious institutions and their power (whether social, political, economic, or symbolic). The second trend is the systematic use by secular authorities of religion as *instrumentum regni*, in which social and political functions (such as the census and the control over subversive elements) were attributed to religious institutions. This process marked a noticeable secularization of religious institutions—something which is usually described as a "degeneration" of the clergy, but which at the same time ensured the highest penetration of Buddhism in Japanese society and culture. Any attempts to redress the situation by reducing the secular role of Buddhism, purifying the clergy, redefining the relations sacred/profane, not to mention anti-Buddhist criticisms and policies were inevitably bound to end up in attacks against the material aspect of Buddhism and thus against its vast system of symbols and representations.[7]

Very few events have had such a dramatic impact on Japanese Buddhism as Oda Nobunaga's attack on Mount Hiei in 1571. Nobunaga had tried repeatedly to subjugate the military power of the Tendai institution but to no avail. He finally decided to get rid of it. His army encircled the mountain and set fire to the forests covering it; he also ordered that all people trying to escape the mountain fire were to be killed. As a result, Enryakuji, Hie Shrine, and hundreds of buildings on the Tendai holy mountain were completely destroyed; thousands of people died; precious doctrinal texts, historical documents, sacred images, and art objects were irretrievably lost.[8] After Nobunaga's death, it took several decades to rebuild Enryakuji, and even then, only on a smaller scale. As an important consequence to this act of destruction, the center of the Tendai sect moved to Edo, specifically to Kan'eiji, one of the most magnificent temples of the eastern capital.

Of course, Nobunaga's attack against Mount Hiei was not an isolated incident, but part of his campaign against the secular power of religious institutions. Between 1570 and 1580 he fought against Ishiyama Honganji, the power center of the Ikkō sect, and destroyed it; in 1583 he arrested and killed more than a 1,300 itinerant priests (*Kōya hijiri*) from Mount Kōya, thereby terminating their centuries-old tradition. Nobunaga's notorious lack of reverence for religious institutions has been emphasized by many scholars. In a famous episode, he signed one of his letters as Dairokuten Maō, that is, Māra, the arch-enemy of Buddhism. But Nobunaga was not completely anti-religious. In fact, he attempted to use religion as a tool for social control; he supported the Christian missionaries (also expecting of them a contribution in his efforts at weakening Buddhism) and more docile Buddhist and Shinto institutions. He also built his own temple, the Sōkenji, as discussed in Chapter 4. In a sense, then, it was Nobunaga who for the first time broke the taboo against a radical rejection of Buddhist institutions in Japan. In his view, religious institutions were not above the secular powers—indeed, there was no place at all for independent religious institutions in the arena of secular politics and power relations; people could believe what they wanted, provided that they subjected themselves completely to secular authorities. After he ordered the total annihilation of Mount Hiei, Nobunaga replied to his retainers' criticism in the following way:

> I am not the destroyer of this monastery. The destroyer of this monastery is the monastery itself [. . .] I have devoted myself to hard work and to a life of denial of my personal desires. I have given myself to the hardships of warrior life in order that I might restrain the turbulence within the land, check the decline of imperial prestige and restore it, improve the prevailing manners and customs, and perpetuate the benefits of government and religion. (Tsunoda, de Bary and Keene (eds), 1964, 1, p. 306)

However, he continued, the soldier-monks of Mount Hiei have constantly tried to thwart his efforts:

> It is they who obstruct the maintenance of law and order in the country. Those who would help the rebels are themselves traitors of the country. If, moreover, they are not destroyed now, they will again become a peril to the nation. Therefore, not a single life should be spared. (ibid.)

In this self-justification, aside from his own moralizing apology, Nobunaga departs from the logic of the mutual dependence of religious and secular institutions discussed above. The "state" (himself) is above religion, and wrong actions by monks are not caused by the interference of "evil spirits"—as Heian period aristocrats asserted; for Nobunaga, such monks are simply "rebels," "traitors of the country," and "a peril to the nation."

Nobunaga was not the first secular authority to carry out a major and devastating attack against a powerful temple. As we have seen before, Taira no Shigehira led an army in an attack to Nara that ended up destroying the Tōdaiji and the Kōfukuji, but at the time it was claimed that such an unfortunate outcome was an involuntary accident. More directly related to our present discussion is the attack on Enryakuji carried out by shogun Ashikaga Yoshinori (1394–1441) in 1433, which ended in the destruction of the Konpon Chūdō hall and the killing of several monks. This attack was also not the consequence of a different vision of the role of religious institutions within a new state order. On the contrary, Yoshinori was still operating according to the medieval logic previously discussed; his attack was an act of retaliation against specific and localized issues, namely, the fact that Enryakuji monks had first burned Onjōji and then cursed him. However, the novelty consists in the fact that it was the highest political authority in Japan, and not a regional or subordinate center of power, that decided to carry out a military attack against a major religious institution. Intriguingly, Yoshinori, before he became shogun, had been the abbot of Shōren'in *monzeki* in Kyoto and even the main abbot (*zasu*) of Tendai at Enryakuji with the name Gien, before he disrobed himself to succeed his brother, the Shogun Yoshimochi, in 1429.

The subsequent history of the relations between religion and the state in Japan essentially followed the path indicated by Nobunaga—albeit without his cruel radicalism. Toyotomi Hideyoshi (1536–98) began the reconstruction of Mount Hiei in 1583, but at the same time he did not hesitate to destroy Negoroji in 1585. Negoroji was the temple founded by the followers of Kakuban when they left Mount Kōya after repeated conflicts with the priests of Kongōbuji. In the late middle ages, Negoroji, with its 2,700 buildings, was one of the major economic and military organizations in Japan, rivaling only Enryakuji and the most powerful daimyos. It controlled an area corresponding to present-day Osaka and part of Wakayama Prefectures, and could mobilize a well-armed and well-trained militia (Negoroshū) of eight to ten thousand men. (Artisans at Negoroji were the first in Japan to build firearms on the basis of a technology stolen from the Europeans in Tanegashima in southern Japan.) Hideyoshi also had an ambivalent religious policy; he supported temples of all denominations, and even promoted the construction of the Hōkōji, as we have seen, a gigantic temple in Kyoto to enshrine the largest Buddha image in Japan, the so-called Kyoto Daibutsu, as part of a project of self-deification.

The third, and most successful, unifier of Japan, Tokugawa Ieyasu, and his descendants, established the strongest regime of premodern Japan, one that lasted for more than 250 years. It is impossible to summarize in a few words the Tokugawa government's religious policies; I will limit myself to mention the definition and codification of each denomination and its teachings; the hierarchical restructuring of religious institutions in head- and branch- temples (respectively, *honji* and *matsuji*); and the obligation for all Japanese to register at a Buddhist temple (*terauke seido*), of which they would become sponsors (*danka*) by performing religious services, in particular funerals. At the same time, religious institutions that were deemed dangerous, such as the Ikkō sect and the radical Fuju Fuse branch of the Nichiren sect, were systematically persecuted in ways similar to those employed against the Christians. With these and other reforms, orthodox institutions became census bureaus for the Tokugawa state, thus coming to play important secular functions related to the surveillance of society (households and villages). I would argue that this secularization of the sacred on an institutional and administrative level constituted a novel aspect of early modern iconoclasm as it attempted a cultural redefinition of Buddhism. It also had significant doctrinal counterparts, as when Buddhist intellectuals began to claim that newly fashionable Confucian values were identical with the Buddhist Law, and to preach that salvation (enlightenment and "becoming a Buddha," *jōbutsu*) could be achieved by carrying out one's secular duties.[9] In other words, the secular became the privileged arena of the sacred, but the sacred was increasingly defined in non-Buddhist terms—mainly Confucian, but in time Nativist, Shinto, and pro-Western movements also developed, thus undermining the intellectual and symbolic grounds of Buddhism. Thus, in the Tokugawa period, Buddhism, despite its permeation of many arenas of the Japanese life and the culture, gradually lost its symbolic power to competing systems of ideas.

Christian iconoclasm

Another series of events that took place toward the end of the sixteenth century also had a lasting impact on subsequent Japanese religious policy. A few regional daimyos who had converted to Christianity had begun, also under advice from European missionaries, to implement violent forms of anti-Buddhism. One case in point is that of Ōmura Sumitada (1533–87), lord of Sonogi, Kyushu. Sumitada was the first daimyo to convert to Christianity; he was baptized in June 1563 as

Bartolomeu. Jesuit Padre Gaspar Coelho convinced Dom Bartolomeu to carry out a forced Christianization of all his subjects; those who refused baptism were exiled. Buddhist temples were turned into churches or destroyed; their priests were encouraged to become Christians by promises to leave them their stipends (Ellison, 1973, p. 92). Another significant case is Takayama Ukon no Suke, aka Justo (1552–1615), perhaps the most famous and influential of the so-called Christian daimyos. Jesuit sources report that he "had destroyed . . . temples and burnt. . . idols" in his domains of Takatsuki first and Akashi later, both located not far from Kyoto (Elison, 1973, p. 125). These acts were unprecedented in Japan. Hideyoshi's edict of 1587 prohibiting Christian proselytizing mentions Christian-directed acts of destruction of religious entities: "To approach the people of our provinces and districts and, making them into [Kirishitan] sectarians, cause them to destroy the shrines of the gods and the temples of the Buddhas is a thing unheard of in previous ages" (quoted in ibid., p. 116). As we have seen, Hideyoshi himself did not hesitate to destroy influential religious institutions, but he was right to stress that until then no one in Japan had attempted to eradicate Buddhism in favor of a new and different religion. In this sense, the "Christian daimyos" were more radically iconoclastic than Nobunaga and Hideyoshi. They did not simply want to reduce the influence of powerful Buddhist institutions and redraw the map of the sacred and the secular in Japan; in addition to all this, they had a new idea of the sacred and its representations, one that was incompatible with previous forms. The Christian rhetoric circulating in Japan at the time was one of the factors at the basis of such new developments. Just to give a few examples, Jesuit Father Valignano claimed that religious perversions had brought God's punishment upon Japan in the forms of war, pestilence, and famine (ibid., p. 41); the Japanese convert Fabian Fucan wrote that Buddhism was the origin of revolt and social perturbation: "Because of the worship of the pernicious Buddhas and gods Japan has been cursed with the punishment of Heaven, so that military conflict has proliferated here more than in any other land" (quoted in ibid., pp. 52–3). Interestingly, however, Fucan later apostatized and wrote a virulent anti-Christian tract entitled *Ha Daiusu* (translated in ibid., pp. 257–91); similarly, Bartolomeu Ōmura, just a few years after his anti-Buddhist iconoclasm, turned into a persecutor of Christianity (ibid., p. 91).

We are not dealing here with a few individuals' loose ethics and confused religiosity; on the contrary, these cases were symptoms of a more general change in religious values. The consolidation of secular power on a regional basis that had begun during the Muromachi period was destined to result in a clash with the secular power of religious institutions. Christianity functioned as a catalyst for a massive campaign against Buddhist institutions aimed at depriving them of their secular possessions and social prestige. Other, more indirect influences can perhaps be found in the knowledge of anti-Buddhist policies implemented in Korea by the Li dynasty from 1392,[10] and a renewed awareness of older, similar instances in China.

As is well known, Christianity was cruelly persecuted in early modern Japan, and the anti-Christian ban was used as a justification for the Tokugawa policies aimed at controlling all religious institutions, including Buddhism. Several feudal domains enforced strict prohibitions against the Ikkō sect and radical Nichiren groups (in particular, the Fuju Fuse sect); a person suspect of belonging to the Ikkō sect might have been arrested, tortured, exiled, or executed.[11] An increasingly strong anti-Buddhist sentiment, which extended to all denominations, began to permeate the entire Japanese society; it was also fueled by the development of a number of competing religious and philosophical discourses, such as Neo-Confucianism, Nativism, Shinto, and modern Western science. The Mito domain carried out a systematic anti-Buddhist iconoclasm in the 1660s, with closure or destruction of nearly 1,100 temples out of the 2,400 existing in the region, and a drastic reduction of the clergy (Ketelaar, 1990, p. 49). The idea underlying the Mito "reforms" was that Buddhism was unproductive if not even antisocial, for it wasted significant economic resources and kept many people (the priesthood) idle. Behind this position we have the Confucian idea that the only acceptable activity is economically productive activity; leisure (often, in premodern times, associated with religious events and festivals) was to be condemned. The Mito religious policies became a paradigm for subsequent anti-Buddhist iconoclasm that took place in the early nineteenth century and, especially, at the beginning of the Meiji Restoration.

Modern iconoclasm

Modern iconoclasm, which begins during the anti-Buddhist persecutions of the early Meiji era and in a sense continues until the present, is of a very special nature. Not many people have been killed—indeed, there has been no strategy aiming to kill Buddhist practitioners. Many artifacts have been destroyed, but as we will see, destruction of certain religious objects was not the primary goal of what was known as *haibutsu kishaku* ("discarding Buddhism and smashing Śākyamuni"). Rather, anti-Buddhist policies attempted a radical re-signification and restructuring of the entire religious field in Japan. A new semiotic system for Buddhism was defined—in opposition to a newly created "Shinto"; religious expressions and representations were limited, codified, and restrained; "traditional" piety was initially forbidden and eventually wiped out in favor of a "modern," "enlightened" religiosity closely tied up with the policies of the newly formed nation state.[12] Perhaps most significantly of all, Buddhist sacred artifacts came to be reconfigured as art objects, state-sanctioned cultural properties; extracted from their traditional cult contexts, they became representations of national identity and "spirit."

At the very beginning of the Meiji Restoration in 1868, the government issued a series of edicts ordering the "separation" of Buddhism from Shinto, the repression of Buddhism, and the establishment of Shinto as the state religion. This operation entailed a definition of "Buddhism" and "Shinto"—what government sources called "clarification" of what was Buddhist and what pertained instead to the *kami* (*shinbutsu hanmei*)—in terms that had little or no basis in traditional religious thought and practice. At the same time, "separation" was carried out in violent and destructive ways: thousands of temples were either destroyed or turned into Shinto shrines or state buildings (schools, army barracks); sacred images and texts were destroyed or sold to merchants; Buddhist funerals and public, festive ceremonies were forbidden, together with other forms of popular religiosity such as Shugendō (mountain asceticism), Onmyōdō (yin-yang calendrical and divination activities), and all kinds of divination, magic, exorcism—even the popular All Souls' Day (*obon*) was outlawed (Ketelaar, 1990, p. 53; Umeda, 1971–2, vol. 3, pp. 15–16).

The destruction of Buddhist temples was more systematic in those feudal domains whose ruling class was heavily influenced by Nativists of the Hirata Atsutane school, such as Satsuma, Tsuwano, Sado, and Fukui. In those regions, the Buddhist presence was practically wiped out. Even in those parts of the Japan in which the separation policy was implemented more leniently, traditional forms of religiosity were almost completely eradicated. In short, the separation of *kami* from the buddhas consisted in the following. Since most Buddhist temples at the time were associated with and controlled a shrine dedicated to the *kami*, these religious complexes were separated; the shrine was made independent, and the deity examined and often redefined. All objects with some Buddhist connotation present in the shrine were discarded; some were destroyed, others still were given to neighboring temples or to lay devotees who preserved them, some others still were sold to antiquarians and dealers in wood and metal scrap. Joint ceremonies (such as sutra readings in front of the *kami*) were abrogated and prohibited. In several cases, the "Buddhist" temple from which the "Shinto" shrine had been separated was vacated, its clergy forced to choose between defrocking or becoming Shinto priests; often, the temple was subsequently destined to a different function as a state building (school, military barracks, stable) or destroyed.

In this way, the religious landscape of Japan changed abruptly. Hie Shrine, enshrining the tutelary deity of Mount Hiei, protector of Tendai Buddhism, was "separated" from Enryakuji and became a completely independent institution (even though joint Shinto-Buddhist ceremonies have been revived). In the sacred landholdings of the Ise shrines there were until 1868 some 110 Buddhist temples; they were destroyed, and even today no temple exists near the shrine. The Hachiman shrine-temple in Usa, northeastern Kyushu, was the place where the god Hachiman (who actually defined himself as "great bodhisattva") originally manifested himself in the eighth century. It was part of a large Buddhist temple, Mirokuji, dating back to the Nara period. Hachimangū itself was governed by Buddhist monks (*shasō*), with *kami* specialists as their subordinates. Mirokuji and its smaller dependencies were dismantled or destroyed, its clergy disrobed or turned into Shinto priests; Hachimangū became a "Shinto" shrine without traces of Buddhism in it. A similar fate occurred to Iwashimizu Hachimangū, located between Kyoto and Osaka. What happened at Suwa, a major cultic complex in central Japan, is also interesting. The two Suwa shrines, Upper and Lower, were controlled by Buddhist temples. The god, Suwa Daimyōjin, was in fact considered a manifestation of Buddhist divinities. The separation edicts were enforced there too, erasing all Buddhist traces from the landscape, the ritual apparatus, and the clergy (Inoue, 2003). (The place where the *bettōji*, the Buddhist temple supervising the Upper Shrine, was situated, is now occupied by a spa–hotel—a magnificent site with views both on the sacred mountain and the lake below.) Equally disastrous was the Meiji policy on Miwa. Situated south of Nara, Miwa had been an important center of Buddhist doctrinal and ritual discourses concerning the *kami* since the middle ages; it was also the seat of a major branch of the Shingon school. Its deity, Miwa Daimyōjin, is Mount Miwa, at the foot of which worship halls and large Buddhist temples, Byōdōji and Daigorinji (also Ōgorinji) existed. Those temples were destroyed, and most of their sacred images and texts dispersed or burned (Antoni, 1995). Total obliteration of the Buddhist presence also took place at Matsuo Shrine in Kyoto. The Buddhist temple (*jingūji*) controlling it, affiliated with the Shingon sect, was completely destroyed; the place where it stood is today a parking lot. The statues of the combinatory gods of the temple/shrine were bought by the Kyoto National Museum, and other artifacts were sold, relocated, or destroyed. One of them, the Buddhist Canon (Daizōkyō) formerly at Matsuo Shrine, was "rediscovered" recently.[13]

However, the zealots who carried out the government's orders sometimes found out that it was not easy to define what pertained to the buddhas and what to the *kami*. The Konpira cultic site in northern Shikoku was a typical example of these difficulties. It enshrined Konpira Daigongen of Mount Zōtō and was governed by a Tendai temple (*bettōji*), the Matsuoji. The priest Yūjō went to Kyoto in 1868 to petition against the implementation of the separation edicts at his religious complex. He argued forcefully that his temple was not a case of mixture of *kami* and buddhas, because its main deity was clearly an Indian god—in fact, it is the Hindu god Kubhera. However, soon afterwards, Yūjō changed his position completely, and argued instead that Kubhera had been assimilated to the preexisting "Shinto" *kami* Ōkuninushi, who should therefore be considered the original deity of the place. Mount Mt. Zōtō was renamed Mt. Kotohira, and the entire cult site took the name Kotohira. Here again the Buddhist temples were destroyed or converted into Shinto shrines, the clergy disrobed or turned into Shinto priests. In this case, besides the abrupt volte-face of the monk (a phenomenon not too uncommon at the time), we see that traditional associations of buddhas and bodhisattvas (worshiped at the *bettōji*), Indian deities (Konpira), and local gods (Ōkuninushi), were clearly difficult to disentangle in order to determine a supposedly original situation before the arrival of Buddhism. In the case of Konpira, it is possible that the mountain had had some religious significance in ancient times, but nothing certain is known about it. We do know, however, that Buddhist priests established there the cult of Kubhera (Konpira), the protector of commerce, travel, fishing, and wealth; it is even possible that it was the Buddhists who sacralized the mountain in the first place, by claiming that a god lived there and that it was a Japanese manifestation of the Indian Kubhera (Thal, 2005).

Similar problems also occurred to cult sites dedicated to the goddess Benzaiten (Benten); these were generally separated from their Buddhist temples and turned into Shinto shrines, despite the fact that Benten is the Indian goddess Sarasvatī. One of the very rare exceptions to the separation policies was the Sōtō Zen temple Myōgonji in Toyokawa near Nagoya, better known as Toyokawa Inari. Since the middle ages this is the center of a cult dedicated to the *kami* Inari envisioned as a manifestation of the Indian goddess Dākiṇī (Dakiniten) and the bodhisattva Kannon (Avalokiteśvara). The priests were able to convince the government that they worshiped exclusively Indian gods, and therefore there was no undesirable mixture of Buddhism and Shinto. The syncretic cult at Myōgonji still continues today (Heine, 1994).

Despite the rhetoric used by government sources and apologists of the separation, with its vocabulary of cultural and spiritual purity, asserting the need to put a stop to religious syncretism and mixture of Japanese and foreign cults, the early Meiji religious policy was a veritable "cultural revolution," in Allan Grapard's definition (Grapard, 1984), that changed forever the religious world of the Japanese. Traditional forms of Japanese Buddhism, in many respects uniquely peculiar to Japan, were destroyed in favor of a sanitized kind of Buddhism separated from local cults and traditions. In addition, Meiji government's Shinto was not the autochthonous, traditional religion of the Japanese, as it was claimed, but a new and artificial concoction, mixing imperial cults, Nativist speculations, Confucian authoritarian ethics, and folklore. Meiji iconoclasm shows important aspects of continuity with early modern iconoclasm, but its originality lies in its radical reformulation of the sacred—in particular, the "Japanese" sacred—and the destruction and negation of everything that did not conform to this new vision.

The early Meiji religious policies were the result of the new government's desire to rid itself of the system of power that characterized the Tokugawa regime, especially its close collaboration with Buddhist institutions, together with the program of the Nativists of the faction founded by Hirata Atsutane, to establish a vaguely conceived Shinto fundamentalist theocracy. The government soon realized the risks of a Nativist theocratic drift and after 1871 greatly reduced their role in policy making. But the cultural damage had already been done, and the peculiar form of Japanese Buddhism, which had integrated Indian, Chinese, and local deities, is now forever lost.[14] The Japanese vision of the sacred and its representations was changed and two entirely new traditions were invented: "Shinto" and "modern Buddhism." As a result, many Buddhist sects responded to the attacks by rejecting much of their own history and taking up the Western project of trying to get back to the "original" Buddhism of the Buddha. They also began to pitch Buddhism to contemporary, Western ideas of religion, which required filtering out anything too "religious" (according to the dominant Western views of the time); the Buddhism that resulted variously became a philosophical psychology, the essence of Japan, the essence of all religions.

In the Meiji iconoclasm, Buddhist (and syncretic) objects were treated as false sacred artifacts to be destroyed as polluting hybrids; often, they were reduced to either "junk," cultural mumbo jumbo to be disposed of, or mere raw materials to be recycled and used in more efficient ways. This is why temples were utilized for different, secular functions, or dismantled and their materials reused, and the clergy disrobed (turned into secularly productive laypeople) or placed themselves at the service of the emperor as Shinto priests. Metal objects such as statues and temple bells were seized by the state to make coins or metal artifacts—mainly weapons, to sustain the prolonged war effort in which the Japanese militaristic governments were engaged between 1868 and 1945. An episode almost forgotten today is the seizure of Buddhist metal implements after 1942. In November 1941, the Dainippon bukkyōkai (the association of Japanese Buddhist organizations), in a spirit of systematic collaboration with the militaristic and authoritarian government of Japan, invited all its affiliated temples owning more than two bells or other metal ritual implements deemed non-essential to donate them to the government. In May of the following year, however, the government issued a formal order (Edict on the special requisition of metal objects, *Kinzokurui tokubetsu kaishū rei*) of seizure of all religious artifacts in metal made after the Edo period and not classified as National Treasures and Important Cultural Properties (Daitō, 1994, pp. 30–4). The rhetoric of emergency was explicitly invoked to justify this state appropriation of Buddhist objects reduced to material resources for the war effort.

In the modern period we also find forms of Buddhist self-damage and self-destruction—but again with a significant twist that distinguishes them from earlier occurrences. The previously mentioned case of the leading monk of the Konpira cult site, Yūjō, who went to Kyoto to protest against the implementation of the separation edicts and ended up by carrying it out himself, is one of self-destruction. In other instances, such as in the Ise Yamada region and at Iwashimizu Hachimangū, priests were given the revenues from the sale of sacred objects they had themselves sold off in the "separation" campaigns (Tsuji, 1984, vol. 4, p. 147). In countless other cult sites, Buddhist monks decided to become Shinto priests in order to save at least a semblance, if not the substance, of their religious tradition. The voluntary donation to the state of metal religious implements during World War II is another case of respected Buddhist institutions carrying out a policy against themselves—their moral and material foundations. At the time, all Buddhist institutions were officially aligned with the government ideology; the donation of precious objects to the state was interpreted as a contribution not just to a secular project, but to a holy endeavor: the Japanese aggression wars in Asia and the Pacific was in fact described as a "holy war" (*seisen*) aiming at the triumph of Imperial Japan's superior civilization and, for the Buddhist organization, of the ideals of Mahayana Buddhism.

To an extent, such attitude was a camouflage to protect Buddhism from a violent and hostile state. However, Buddhist institutions' explicit subordination to the emperor and the state was a dramatic reversal of medieval ideas of mutual dependence. Now it was no longer the emperor and the state that supported Buddhist organizations and organized fundraisings to promote Buddhism, as in the past; on the contrary, it was Buddhist temples that raised funds for the state and its wars of aggression. For example, in November 1942 the Ōtani branch of Jōdo Shinshū even launched a fundraising campaign to support the construction of warships and warplanes (Daitō, 1994, pp. 34–6). In this respect, we could also understand as a case of self-destruction the Buddhist institutions' organic support of Japanese militaristic imperialism, with the authorship of what Ajiki Bun'yū called "apocryphal scriptures," which mobilized Buddhist vocabulary and images to justify war and aggression (Ajiki, 2002, pp. 31–7). As Christopher Ives has written,

> From the 1920s to the early 1940s Buddhist priests contributed actively to Japanese imperialism. They lent their social status and homiletical skills to propaganda campaigns (*kyōka undō*) run by the state to cultivate obedient imperial subjects; organized and/or participated actively in patriotic groups; exhorted parishioners to "serve the public" (*hōkō*) by enlisting, practicing austerity on the home front, and buying war bonds; engaged in monthly "patriotic alm-begging" (*hōtoku takuhatsu gongyō*); donated temple funds for the construction of warplanes; ran officer-training programs; performed ceremonies and chanted sutras to promote Japanese victory; assisted the families of the war dead; served as chaplains for troops fighting overseas; and helped "pacify" (*senbu*) occupied and colonized areas and mold colonized Asians into imperial subjects (*kōminka*). (Ives, 2001, p. 16)[15]

Among the arguments of Buddhist Imperial propaganda, we find the following:

> 1) the war Japan was waging across Asia was a holy war; 2) Japanese actions in that war were expressions of compassion; 3) the deaths of brave, self-sacrificial soldiers embodied the Buddhist doctrine of no-self; 4) through this Buddhistic self-sacrifice Japanese could repay their debt (*on*) to the emperor; 5) the Japanese military was attempting to establish a pure land here on earth, with the emperor equivalent to Amida. (Ives, 2001, pp. 16–17; see also Ives 1999, 2009)

Differently from medieval self-damaging actions, which were essentially strategies employed by Buddhist institutions to display and emphasize their own importance, modern self-damage painfully indicated the superfluity of Buddhism for the state and its impotence against state power.

To conclude our discussion thus far, we could say that there has been a substantial continuity between early modern and modern *forms* of iconoclasm, insofar as sacredness was radically denied and sacred objects and the bodies of religious specialists were reduced to raw materials and raw physical force to fuel productive activities at the service of the state and the secular authorities in general. However, early-modern and modern forms of iconoclasm were radically different in their *significance*, in particular in the scale of the attempt to deny the sacredness of Buddhist representations that occurred in modern Japan which was, as we have seen, unprecedented and had consequences lasting until today.

Another, very important effect of modern iconoclasm is the conversion of religious artifacts into art objects. Already in 1876 aesthetic considerations, in particular the idea that at least certain temples represent Japan's "concept of beauty" (*bikan*), were adduced as reasons not to destroy them; appreciation by Westerners was an additional incentive (Tsuji, 1984, vol. 4, pp. 150, 151). In this way, modern iconoclasm found its definitive success: as we will see in Chapter 4, preservation of religious artifacts as art objects saves the materiality of the object but destroys its sacred significance.

Issues of sacredness, however, are predictably not so simple and clear-cut. In the course of these developments many Buddhist artifacts have come to be considered important "art objects" exemplifying the traditional artistic sensibility of the Japanese, and this newly acquired status has undoubtedly contributed to the fame and prosperity of temples enshrining them. Temples thus advertise the display of their sacred objects (images, implements, etc.) with the hope of drawing attention and receiving more visitors. Yet we have a paradox here. While the advertisement of a religious object as an art treasure participates in its desacralization, recognition of an icon as an art object may actually serve also to strengthen its sacred value and even contribute to the "prosperity of Buddhism"—or, at least, of a certain temple. Sacred objects, thus, are not only part of an ongoing process of commodification of the sacred; they also play an important role in a parallel process of resacralization. Embodying the mysterious power of the Buddha, they further stand as relics from the past, from the "authentic" Japanese tradition.[16] And as such, they can bring to the temple tourists and new potential patrons (*danka*). It would seem that the desacralizing logic of modernity, that which produced "art objects" out of sacred icons, was not completely successful in this case. After all, it is not unusual to see people joining their hands and bowing to a particularly famous buddha image on display at a museum exhibition; tourists visiting famous temples in Kyoto and Nara often pray there. Thus, the boundary between art objects and cult objects may not be as clear-cut as many of us would like to think. Perhaps, one might argue, what has really been at work is a counter process of resacralization of the images as embodiments of the national spirit, as state icons, through exploitation of what Michael Taussig (1997) calls the "magic of the state," as we discuss in Chapters 4 and 5.

To conclude this section, we should mention a very special case of modern iconoclasm, involving a complex and conflictual relationship between the religious and the aesthetic, namely, the burning of the Kinkakuji in 1950 by a psychologically unstable young monk. As narrated by Mishima Yukio in his novel *Kinkakuji*, the temple became the object of the protagonist's paranoid obsession with beauty, impotence, nothingness, and death (Mishima, 1959). The protagonist of Mishima's novel, by destroying the Kinkakuji, was also making a powerful statement about his own outlook of life; he was challenging tradition and promoting revolutionary change, albeit in a very confused way. As such, the burning of the Kinkakuji goes beyond an act of vandalism by a deranged individual; the beauty of the temple became the target of the destroyer's obsession and his doomed revolutionary statement.

It should be clear from my discussion so far that iconoclasm does not simply attack material objects or the physical bodies of entities and people representing/embodying the sacred. On the contrary, iconoclasm is always also a semiotic struggle aimed at re-signification and re-configuring of the sacred and thus the relationships between sacred and profane. The meaning of destruction was also complex and manifold, varying according to the perpetrators, the recipients, and the observers now and then, here and over there. It is necessary, then, to outline a hermeneutics of iconoclasm, in order to understand in what terms destruction was explained, justified, rationalized, or criticized.

History as a cosmic state of siege

What were the reactions of premodern Japanese to the destruction of Buddhist temples and other important sacred artifacts? Considering the sacred status of temples and the flow of destruction underlying the history of Japanese Buddhism, their destruction must have provoked strong sentiments of various kinds.

Temples and sacred artifacts in general, were envisioned as "power spots" where the visible and the invisible intersected; they were permeated with supernatural power, and that power could not be destroyed. A tale from *Uji shūi monogatari* (early thirteenth century), entitled "The Holy Man of Shinano Province," tells of the sacred power that lingers on a religious structure—a temple's storehouse, the subject of a famous miracle story—even after its destruction:

> The storehouse is now in a very dilapidated state, but is still there. People who manage to get hold of even a splinter of the wood from which it was built, make it into a charm, and those who obtain larger pieces and have them

> carved into images of Bishamon never fail to become rich and prosperous. It is no wonder that everyone who hears this story is anxious to buy a piece of wood from the storehouse. (Keene (ed.), 1955, p. 223)

In the same way, a whole narrative genre existed, consisting of tales recounting that sacred images should not be destroyed, also because it is in fact impossible to destroy them. These miraculous stories tell of sacred artifacts such as icons and holy texts that either survived fire and other agents of destruction or were just temporarily reduced to ashes, only to subsequently transform themselves into other sacred artifacts—for instance, a hanging scroll with the formula Namu Amida Butsu inscribed on it was burned, but its ashes turned into a "golden Buddha" image (Daiki, 1995, p. 4). If sacred artifacts possessed a spiritual power that survived, or at least resisted, destruction, as according to the Buddhist rhetoric of animation discussed in Chapter 1, their disposal should also be ritually controlled. In fact, images that had been damaged beyond repair were treated in several ways, all respectful of their intrinsic sacredness. For instance, they could be stored away somewhere in a temple, as in the case of the Buddha head from Kōfukuji (see Rambelli, 2001), or recycled—melted down—to produce a new image, as in the case of bronze statues. When complete disposal was necessary, wooden images were burned, but the symbolic imagery that was mobilized in the occasion emphasized this act's similarity with Śākyamuni's cremation or with forms of self-sacrifice, as in the case of icons becoming fuel for a temple's public bath (such as Rennyo's Kudoku-yu) (see Daiki, 1995, pp. 8–12).

In general, however, sources report expressions of utter disquiet and puzzlement when they describe people's feelings upon hearing of or assisting in the destruction of a major temple. Authors decried the event as "inexplicable" or even "disgusting";[17] even the deities, claimed one source, must have been "struck with awe at the sight of this terrible disaster." At times, the scale of destruction was so staggering that deities and even nature manifested signs of grief. The *Tale of the Heike* concludes the description of the destruction of Tōdaiji and Kōfukuji by the Taira army in 1180 with the following words:

> For years afterwards the hearts of men were heavy with sorrow. Bonten, Taishaku, the Kings of the Four Quarters, the Eight Guardian Gods, and even the jailors and demons of hell must have been struck with awe at the sight of this terrible disaster. And Kasuga Daimyōjin, the guardian god of the Hossō sect—what must he have felt? Even the dew on the Kasuga Plain changed its hue, and the wind from Mount Mikasa moaned. (Kitagawa & Tsuchida, trans., 1975, vol. 1, pp. 342–3 with minor changes)

Here the disaster transcends the time and place in which it happened and assumes cosmic dimensions; the destroyed artifacts became monuments to universal decay and the presence of evil in this world.

Nevertheless, destruction was envisioned as an event that did not necessarily threaten the sanctity of a temple; on the contrary, as a sign, as a meaningful act of communication from the invisible world, destruction actually reinforced sanctity. Even when destruction was apparently unexplainable, exegetes maintained that it was not an absurd event, but that there was a hidden logic beyond human understanding. This hidden logic was defined as *fukashigi*, "beyond comprehension," an epistemological category usually applied to the secret intentions of buddhas and *kami* and to the deepest features of enlightenment. In those cases, sources often choose the rhetoric strategy to indicate that people could not make sense of what had happened and that language was unable to represent their emotions. After the burning of Kiyomizudera, as we have already seen, another notice replied that the power of Kannon, Kiyomizudera's main image, is "eternal and beyond the reach of human knowledge. None can fathom the reason why Kiyomizu burned down" (Kitagawa & Tsuchida (trans.), 1975, vol. 1, p. 41). In other words, disbelief, or at least suspension of belief concerning the sacredness of a temple was countered by emphasis on the incomprehensible, but ultimately benign, intention of the deity involved. This attempt to rationalize the inexplicable is not dissimilar to claiming that, in the words of philosopher George Santayana famously quoted by Clifford Geertz, "God is not mad" (1973, p. 99). According to this logic, what appeared as mere natural catastrophes, without any clear message from the Invisible, could be explained as attempts at communication. As Geertz put it, religion serves to give a sense to apparent meaninglessness.

But what kinds of messages were conveyed by acts of destruction against sacred artifacts? The answer was always dependent upon the interpreters and the historical situation. As Mikael Adolphson has written, "It was not uncommon to explain disasters, both personal tragedies and human suffering caused by nature, as the native gods' response to 'political injustices,' such as the neglect of one of the elite temples" (Adolphson, 2000, p. 267; see also McMullin, 1988, p. 272). However, premodern Japanese belief in supernatural intervention was not limited to a few major temples and their political interests; it was instead doctrinally based on a particular interpretation of Buddhist cosmology.

Modern accounts usually explain away temple destruction either as accidental or natural occurrences, or as essentially political acts motivated by power struggles and not by theological/philosophical discussions on the sacred nature of the objects destroyed. Premodern Buddhists, however, saw things in a rather different way. It was important for Buddhist institutions to find explanations that could make sense of destruction in general terms without renouncing the possibility of carrying it out themselves against their religious enemies. Sources tended to collapse the distinction between natural and intentional causes by claiming that all forms of destruction were due to the intentional intervention in this world of beings from the Invisible world of buddhas and *kami*. For example, a fire in Kyoto in 1177 caused enormous damage and many people died. According to the *Tale of the Heike*, that fire

> was no natural event. It must have been the god of Mount Hiei who had dealt the capital this blow, for someone in a dream saw two or three thousand large torch-bearing monkeys coming down from Mount Hiei and setting fire to the capital. (Kitagawa & Tsuchida (trans.), 1975, vol. 1, pp. 73–4)

The destruction of the Zenkōji Amida triad later in the same year was considered "particularly ominous" because of the importance and history of the icons. As the *Tale of the Heike* reports, "'The destruction of Buddha's Law foreshadows the destruction of the Imperial Law,' people said. 'The destruction of many holy temples and mountains portends the fall of the Heike'" (ibid., p. 140). Later, after the defeat of the Heike clan in 1185, a great earthquake that devastated Kyoto was also interpreted as an intentional occurrence; in particular, "intellectuals lamented that the earthquake had been caused by the evil spirits" of the deceased emperor Antoku and the defeated Heike warriors (ibid., vol. 2, p. 720). Even wear-and-tear was sometimes attributed to the will of supernatural beings. Raigō (1002–84), the abbot of Onjōji who struggled to establish an ordination platform at his temple in order to render it independent from Enryakuji, died without accomplishing the endeavor to which he dedicated his entire life. For this reason, medieval sources attribute him with revengeful intentions against Mount Hiei; some texts even describe Raigō as willingly reborn as a rat living in the scripture storehouse on Mount Hiei, where he and his companions would literally gnaw away the most precious documents of that rival cultic center.[18] Stories like this one have an ambivalent content: Raigō, defeated and full of resentment, is reborn as a vile animal, but is able to exploit the power derived from his weak position—even as a rat (or perhaps, as a rat even more than as a human being) he is able to damage Enryakuji.

In this way, Buddhism provided a vocabulary and a set of metaphors to describe and "domesticate" cases of destruction, as the fights among the Asura,[19] or as the cosmic struggle opposing the Buddha and Māra. In the latter, the supernatural enemies of Buddhism, led by Māra, were trying to destroy the very presence of Buddhism in Japan, while supernatural allies of Buddhism were using destruction to send signals and warnings to the faithful, urging them to proliferate their acts of devotion. Indeed, the idea of the intervention of the invisible realm in this world's occurrences is a *topos* running through premodern Buddhist literature. A text written in 1136 by the Shingon monk Kakuban (1095–1143) presents natural disasters such as earthquakes as echoes in this world of the cosmic struggle taking place in the invisible realm between buddhas and Māra's followers.[20] The *Taiheiki* reports two famous episodes explicitly addressing this theme. In the first, the monk Jōkei (1155–1212) once had a revelatory dream at Ise's Outer Shrine. He found himself in the Sixth Heaven, among supernatural beings (demons and asuras) who were complaining about his own (Jōkei's) success in spreading Buddhism; they thought it was thus necessary to confound his heart. Then, Māra himself spoke, revealing his plans. He was to cause a struggle between retired emperor Go-Toba and the Hōjō family (the regents of the Kamakura military government); as a result, the imperial army would be defeated and the emperor exiled. (This was in fact what happened in the Jōkyū Disturbance of 1221.) The shogun Hōjō Yoshitoki (1163–1224) would then appoint as the new emperor Go-Horikawa (1212–34), who will in turn appoint Jōkei a court monk. Busy with ceremonies and politics, Māra continued, Jōkei "will become more neglectful . . . and his pride will increase constantly until he becomes a rule-breaking, shameless monk. Likewise, in this way we shall gain new followers" (McCullough (trans.), 1959, p. 370). A history text such as the *Masukagami* (ca. 1368–1376) also suggests that the Jōkyū Disturbance had supernatural causes: "We must look beyond this world for an explanation of what happened—a truth incomprehensible to those too ignorant to understand karmic law" (Perkins (trans.), 1998, p. 54); the text suggests that it might be due to the anger of Sannō, the god of Mount Hiei (ibid., p. 52). It is interesting that *Masukagami* and *Taiheiki* agree with the idea that occurrences in this world are caused by events taking place in the invisible realm of deities, but disagree on which deities cause what and their motifs. For the *Masukagami*, natural and social alterations were caused by deities aiming at strengthening Buddhism, whereas for the *Taiheiki* it was Māra and his demons who were erecting obstacles to Buddhism. The second episode in the *Taiheiki* dealing with demons recounts of a secret meeting on Mount Atago, northwest of Kyoto, well known as a residence of *tengu* (goblins), held by influential demons ("Great Māra Kings," as the text calls them), who in their previous lives were emperors (Go-Toba, Go-Daigo) and high-ranking priests (such as Genbō,

Shinzei, and Raigō), to conspire about how to cause troubles in the realm (Gotō and Kamada (eds), 1960–2, vol. 3, pp. 59–60).

We find here an interpretive difference on the meaning of bad things: do they happen as hindrances to the prosperity of Buddhism or rather in order to promote it? This difference can be overcome if we consider the medieval esoteric doctrine of the identity of Māra and the Buddha (Jp. *Ma-Butsu ichinyo*). According to it, the Buddha and Māra are in fact cooperating, each in his own capacity, to further a soteriological project. A particular instance of this cooperative effort is the strange collaboration between Moriya and Prince Shōtoku mentioned in medieval texts (see Rambelli, 2010b). These attempts to explain dramatic occurrences in this world, including large-scale destruction of sacred objects, find their roots in the Buddhist philosophy of history.

The most complex and systematic explanation of the Japanese Buddhist philosophy of history was proposed by the Tendai scholar-monk Jien (1155–1225). Jien envisioned the history of Japan as a local manifestation of universal principle (*dōri*)—essentially, a law of cosmic degeneration. Consistent with the scriptures, Jien was connecting the process of self-neutralization of Buddhism (the concept of the end of the Dharma, *mappō*) with cosmic decay. He is the author of a famous poem describing the process (Perkins (trans.), 1998, p. 40):

> Because the time had arrived
> When the holy Law
> Was destined to disappear
> Like froth on a stream.

The idea of degeneration of Buddhism was associated with images of decay of its material substratum, that is, temples, images, and texts. A passage from the *Tale of the Heike* is representative of this sensibility:

> Now that the world has become secular and degenerate, the Buddha's Law in the three countries India, China, and Japan has gradually declined. If we examine the remains of Buddhism in far-off India, we see that the Grove Monastery and the Gion Temple, where of old the Law was preached by the Buddha, are now the haunts of wolves and foxes. It is said that there remains nothing but the foundation stone of those monasteries [. . .] In China too T'ien-t'ai-shan, Wu-t'ai-shan, Pai-ma-ssū, and Yü-ch'uan-ssū now lie in ruins. The sacred volumes of the Mahayana and Hinayana are rotting away in the bottom of their boxes.

> Turning back to our own country, we see that the seven great temples of Nara are now deserted, and so the eight or nine sects have vanished without a trace. In olden days, at Atago and Takao, the temple halls and pagodas lifted their tall roofs in a long range. But in one brief night they were ruined and became dwelling places for *tengu*.
>
> On the mountain [Hiei] now . . . may it not be that the noble law of the Tendai sect was likewise being abandoned? How is it possible that this sect that had prospered from ancient times was now falling into decay? (Kitagawa & Tsuchida, trans., 1975, vol. 1, pp. 137–8)

This passage presents destruction as a natural and inevitable occurrence, rather than something caused by human intervention or carelessness.

However, Jien did not have in mind a mechanistic model of universal destruction, but allowed instead for the possibility of local interventions to counter and delay the effects of the underlying universal principle of decay. He wrote: "deterioration, occurring spontaneously and naturally (*hōni*), cannot be checked by the power of man. But in Buddhist Law there is a way to partially negate (*taiji no hō*) it" (Delmer & Ishida (trans.), 1979, p. 35), for "deterioration alternates with improvement" (ibid., p. 36). Specifically, Jien describes a complex interaction between two different orders of reality: the visible (*ken*, the world in which we are) and the invisible (*myō*, the realm of supernatural beings). He argued that when *kami* or buddhas see a problem in the maintenance of the visible order they manifest an ominous sign—in particular, a disaster affecting temples. The goal of these signs is to show humans the natural tendency toward degeneration so that they can take countermeasures. In some cases, destruction was the consequence of struggles taking place within the Invisible realm between deities and evil spirits—probably an echo of the ancient Indian cosmology opposing gods (*deva*) and anti-gods (*asura*). There is no distinction here between natural and intentional causes of destruction: the ultimate rationale was to be found in the invisible world, and the agency that manifested the sacred signs (human action or natural disasters) was not very relevant.

In fact, one of the most powerful defenses against cosmic degeneration was to support religious institutions. Temples that had been destroyed must be rebuilt; new temples must be built; the Buddhist teachings must be spread, and ritual action must be intensified on all levels. In general, the doctrines of the end of Buddhism (*mappō*) did not produce fatalism and resignation, but resulted in an increase in religious activities and in the strengthening of religious institutions (Hayami, 1991). Thus, destruction was explicitly used to strengthen the role of temples. As Janet Goodwin has written,

> In the world that the monks imagined, temples were not only vehicles for spreading the Buddhist law, they became the Buddhist law itself; thus when a temple was preserved, restored, or constructed, the Buddhist law was as well. Temples—and bridges too, since there was no clear line between sacred and secular worlds—were part of the physical framework in which the Buddhist law was realized and would continue to be realized even in the darkest days of *mappō*. Their construction, moreover, was seen as an act of salvation for all who participated in it, whether solicitor, carpenter, or donor—making it necessary to invite broad participation in the form of donations to *kanjin* [religious fundraising] campaigns. *Hongaku* [original enlightenment] teachings allied Nara monks, vagabond *hijiri* [itinerant holy men], and lay believers, and made all three the instruments of religious change. (1994, p. 145)

In general, however, religious institutions' material integrity and prosperity were signs of order and harmony (on both social and cosmic levels), while their decline was a sign of oncoming chaos and need for countermeasures. When large-scale destruction could not be prevented or stopped, authors thus tried to rationalize cases of destruction as forms of Buddhist institutions' self-sacrifice—self-sacrifice that was envisioned, in turn, as a way to deal with supernatural signs from supernatural agencies.

Whenever a temple was destroyed, its status as a sacred site and the legitimacy and morality of its clergy were questioned, sometimes openly. The case of natural/supernatural destruction was, as we have seen, more nuanced and open to interpretation, but to human destroyers, the temple was not a real temple, and they were eliminating a source of counterfeit, evil Buddhism. Buddhist doctrines state explicitly that, as we have seen in Chapter 1, corrupt monks and heretics together will bring about the End of Dharma and the end of the world. According to this Buddhist logic, the destroyers claimed to reestablish socio-cosmic order and to protect the state and the people by purifying Buddhist institutions of their evils. The destroyers, whoever they were, were more or less explicitly taking upon themselves the duty of the secular government as the protector (and supervisor) of the Samgha. The Taira generals, Oda Nobunaga, Toyotomi Hideyoshi, and to certain extent also the Meiji bureaucrats, who launched massive iconoclastic campaigns, claimed (or could claim) to be acting in order to purify a degenerate Buddhist clergy. In other words, their anti-Buddhist actions could be legitimate and justified both doctrinally and politically.

From the Middle ages on, however, Buddhist institutions also developed a different interpretation. They tried to justify particularly serious and devastating occurrences as forms of voluntary self-sacrifice performed in accordance with the will of buddhas and *kami*. One of the first instances of this logic is an explanation of the Mongol Invasions reported in the *Nomori kagami*:

> Since we are in the final age of the Dharma, the power of Buddhism is declining [. . .] perhaps it was the gods, who, in order to revive the power of Buddhism . . . provoked the foreigners' attack as a skilful means and brandished their sacred swords? (Hanawa and Ōta, eds, 1959–60, vol. 27, p. 508)

In other words, it was the Japanese gods in charge of protecting the country who actually unleashed the Mongols' attacks, which were nothing else than expedient means (*upāya*) to "revive the power of Buddhism." This was by no means an isolated position. Violent means were necessary to save the Japanese. The *Shasekishū*, a collection of Buddhist tales written in the early fourteenth century by Mujū Ichien (1227–1312), is even more explicit:

> Ours is a country as remote from this center [India] as the small, scattered millet seed, where rough, fierce creatures were unaware of moral causation. For those who did not believe in the Dharma [the Buddha in its absolute and unconditioned form] employed that which was appropriate to the time and place. Manifesting the shapes of evil demons and wicked spirits and showing forth the forms of poisonous serpents and fierce beasts, it subdued this ferocious and evil lot and thereby brought people to the Way of the Buddha (Morrell (trans.), 1985, p. 78).

Deities protecting Japan are actually "demons and dangerous animals" or, more literally, "evil demons, malignant *kami*, poisonous snakes and violent beasts"; several Shinto texts belonging to the Buddhist tradition make clear that the true shape of the *kami* is that of snakes. Thus, Japan was protected by dangerous entities against the evil influences of demons.

According to this reasoning, iconoclastic attacks against Buddhism were in fact the *compassionate* deeds of buddhas and *kami*. One of the first instances of justification of anti-Buddhist destruction by Buddhists was proposed in favor of Taira no Kiyomori, the warrior who staged a coup and tried to restructure the relations between the state and Buddhist organizations; he ordered the attack on Tōdaiji and the Kōfukuji that resulted in many deaths and the destruction of the two temples. The *Tale of the Heike* claims that, despite the fact that many people thought Taira no Kiyomori an evil man, "there was in Kiyomori something supernatural that suggested that he was an incarnation of a god or a Buddha";

indeed, he was "in truth an incarnation of Bishop Jie" (Kitagawa & Tsuchida (trans.), 1975, vol. 1, pp. 372, 373). Jie Daishi, the honorific appellation of the Tendai leader Ryogen (912–85), was believed to have become a *tengu* (goblin), known as Ganzan Daishi or Tsuno Daishi, a supernatural entity who vowed to protect the Tendai institutions in Japan. The *Tale* further reports an anecdote concerning a scholar-monk named Jishinbō Son'e. In a dream he was summoned to the court of Enma, the king of hell, who told him that Kiyomori "is not an ordinary man—he is the reincarnation of Bishop Jie. He was born again in Japan in order to protect the law of Tendai Buddhism" (ibid., p. 376). Enma gave Son'e a poem he had written in admiration of Kiyomori (ibid.):

> I revere the great bishop Jie,
> Protector of the Tendai Law.
> You have appeared again as a great general;
> Your evil deeds have inspired men to virtue.

Son'e read this poem to Kiyomori, who was predictably "greatly pleased" and gave the monk gifts and a promotion (ibid.). It is possible that Son'e was an opportunistic sycophant; however, his (or rather, King Enma's) interpretation of Kiyomori's actions were doctrinally legitimate as a case of "inverse karmic relation" (*gyakuen*) with Buddhism, according to which evil deeds can become the inspiration for good actions.

A similar line of reasoning was employed during the early Meiji period to justify Oda Nobunaga's destruction of Enryakuji. The Buddhist author Ashitsu Jitsunen argued that "Hiei merely used Nobunaga to purify itself" (quoted in Ketelaar, 1990, p. 164), thus giving another spin to the previously quoted Nobunaga's own disclaimer (in practice, "I didn't do it; the monks asked for it"). The destroyers of Buddhist institutions were in fact protecting Buddhism from itself! Destruction had become an act of self-sacrifice and self-purification.

Finally, we cannot ignore the fact that premodern sources also include skeptical and non-religious interpretations of occurrences of natural destruction—accounts that question, implicitly or explicitly, the actual role of divinities, if not even the sacredness of the objects affected, as when monks from Enryakuji, after their burning of Kiyomizudera, left a notice questioning the sacredness of the temple. Another example is particularly revealing of more general attitudes. The 1177 great Kyoto fire, that the *Tale of the Heike* considered, as we have seen, as a punishment from the god of Mt. Hiei, also received a different interpretation as the unforeseeable result of a smaller fire that got out of control. For example,

Hōjōki by Kamo no Chōmei (1155–1216), who witnessed the great fire in person, stresses the waste of resources and the ultimate pointlessness of "expending treasures and spirit to build houses in so dangerous a place as the capital" (Keene (ed.), 1955, p. 199). Chōmei also used this natural disaster for political criticism (Marra, 1991, p. 84). Thus, not all instances of destruction were always and necessarily manifestations of divine intervention. Often, Buddhist authors show a co-presence of a secular, empirical attitude, with a cosmic, meta-historical vision involving the invisible world of the deities. In 1602, during the casting of the bronze for the reconstruction of the Kyoto Great Buddha, originally built by Toyotomi Hideyoshi, a fire accidentally spread to the building, which was reduced to ashes. Gien (1558–1626), the head priest (*zasu*) of Daigoji, wrote that this disaster was due to the choice of cheap materials and a below-standard construction method; however, he also added that this was an action perpetrated by Māra (Tenma), a sign of decadence of Buddhism (Murayama, 2003, p. 116).

Conclusion: Managing chaos, or culture and violence

In order to better comprehend the Japanese Buddhist understanding of iconoclasm and acts of religious destruction in general, we may recall here Benjamin's notion of history as a state of siege. As Michael Taussig explains, this position "recognizes that the terror in such disruption is no less that that of the order it is bent on eliminating" (1992, p. 13):

> in a state of siege order is frozen, yet disorder boils beneath the surface. Like a giant spring slowly compressed and ready to burst at any moment, immense tension lies in strange repose [. . .] if we are to take the full measure of Benjamin's point, that the state of siege is not the exception but the rule, then we are required to rethink our notions of order, of center and base, and of certainty too [. . .] our very forms and means of representations are under siege. (ibid., p. 10)

Jien's philosophy of history, based on scriptural sources on the End of Dharma, was a tragic cosmic vision that became the basis for direct engagement in society. The temples were semiotic apparatuses that gauged messages from the Invisible and struggled continuously to prevent the collapse of order. Temples (and religious institutions) were always vulnerable to destruction; but by explaining it in terms of interventions of the Invisible, they required constant attention to the Invisible in order to prevent violence; in other words, Buddhism was in a constant "state of emergency": politically, socially, cosmologically. A large amount of ritual activity (and labor and wealth) was dedicated to pacifying potentially threatening sacred beings, which were therefore constantly evoked and, as it were, always present. And it is precisely this menacing presence that Buddhist institutions exploited: funerals provided salvation, but also prevented the dead from turning into evil spirits; rituals for the protection of the state averted possible calamities caused by deities and demons; transmission of the precepts encouraged controlled morality and therefore decreased the power of the violent Outside. After all, "salvation" and "liberation" can be predicated only upon ideas of damnation and bondage. In premodern Japan, the temples did not try to hide their dramatic history. In Taussig's words, they did "not aim at destroying memory. Far from it. What [they] aim[ed] at [was] the *relocation and refunctioning of collective memory*"; in particular, they deployed "the strategic art of abnormalizing" (ibid., p. 48).[21] Behind a screen of normalcy and peace there was the threatening presence of chaos and destruction. Both secular and religious institutions, and the whole of society for that matter, were built upon a delicate balance—on the verge of chaos, as it were. In modern Japan, all this has largely disappeared, probably because temples play a much more marginal function in society and are no longer the guardians of socio-cosmic order. The history of the destruction of temples indicates to us the seriousness with which these notions were taken, both by the destroyers and the religious institutions, and the manifold intellectual ramifications of actions that, while appearing as mere brute violence, reveal instead deep intellectual and cultural elements.

In principle, destruction threatened the temples' claims to sanctity and challenged people to find valid explanations and rationalizations to those fateful events. We can identify two general trends in the Buddhist rationalizations of religious destruction. The first trend is that natural occurrences of destruction were understood as being in fact intentional acts of destruction, as signs from the invisible realm of buddhas and *kami*. The second trend is that anti-Buddhist violence was reinterpreted either as Buddhist purification, necessary for self-improvement, or as Buddhist self-sacrifice, necessary for the preservation of the cosmic, social, and moral orders. According to this justificatory logic, destruction called for reconstruction (in turn, associated with diffusion and prosperity of Buddhism); violence indicated importance, relevance, and respectability of Buddhist institutions; and social criticisms against Buddhist were neutralized as necessary steps in a process of self-purification. In extreme cases, having undergone acts of destruction could even prevent Buddhist institutions from significant moral reform, since self-purification and thus moral improvement were envisioned as automatically resulting from material destruction.

Thus, destruction could be presented not as an intentional anti-Buddhist act, for the destroyers were actually being manipulated by the forces of the invisible realm that were sending messages to the humans. Natural disasters could be explained in the same way. In some cases, destruction was the direct consequence of struggles between Buddha and Māra taking place in the invisible world, which gave the temples the opportunity to participate in it on the side of Buddha (order and harmony). Moreover, the theme of "degeneracy of the clergy," once understood as the cause of destruction of a temple, could also become a cause for redemption. In this way, destruction was re-envisioned as an essential step toward regeneration. The dreadful effect of raging fire on temples was thus very close, symbolically, to the purifying fire of *goma* ceremonies; and the destruction of a temple was, in turn, symbolically close to individual actions of self-sacrifice for collective salvation.

By strengthening their mastery over signification and their role as interpreters of the hidden mechanisms of reality, Buddhist institutions transformed themselves from passive subjects of destruction into active interpreters of destruction. Destruction thus domesticated did not diminish the power of religious institutions; on the contrary, religious institutions often benefited—symbolically, politically, and economically—from acts of destruction against them. They were often able to capitalize on their temples' destruction by launching massive fund-raising campaigns that boosted the temple economy and resulted in a broad diffusion of Buddhism, as in the case of the reconstruction of the Tōdaiji after 1180. The rulers of premodern Japan generally contributed lavishly to the reconstruction of Buddhist temples and sacred objects, also as a way to reinforce mutual support. In this way, destruction could become a key ideological and political tool used to reinforce the importance of religious institutions, in an interesting symbolic dynamics connecting disaster and prestige. In modern times, after the Meiji-era anti-Buddhist persecutions, which succeeded in part in de-sacralizing Buddhist artifacts, Buddhist institutions have attempted to re-sacralize themselves by connecting the religious "soul" (*tamashii*) of the Buddha in their icons to the aesthetic spirit (*kokoro*) of Japanese culture.

3

Shattered on the Rock of Ages: Western Iconoclasm and Chinese Modernity

Eric Reinders

Early in this book we asserted that the study of iconoclasm should not be limited to the particular cases with which it is commonly associated, namely the Byzantine and Reformation iconoclasms. Throughout most of history, the destruction of sacred objects, and its attendant discourses, developed in East Asia with no reference to Christianity. But beginning in the sixteenth century, Christian missionaries brought to East Asia not only truths to be asserted but also truths to be denied, images to be made and unmade. There is a "negative" history of missions yet to be fully written. It appears that the Chinese connected their own anti-religious rhetoric to new ideas imported from the West, in a process that fueled modern iconoclasm. This chapter outlines Christian iconoclasm in China, and Chinese (especially Communist) anti-religion in the twentieth century.

The Catholic missionaries of the Ming and early Qing envisioned the images of the Chinese and Japanese as idols, and accordingly instigated acts of iconoclasm. Meanwhile, Catholicism in Europe was subject to harsh iconoclastic criticism of its own supposed idolatry, so the Jesuits were careful to distinguish between their own obeisance to images and the obeisance to pagan idols (Reinders, 1997). The elite Chinese on whom the Jesuits focused their attention were often in agreement that popular worship was superstitious idolatry (though they could not have used those terms yet), and that the ancestral cult could be understood in relatively humanistic or psychological terms. By contrast, other Catholic missionaries, and, in the nineteenth century, both Protestant and Catholic missionaries, were far more concerned with converting the masses from the ground up. In general, Protestants saw the "worship" of ancestors and of Confucius without benefit of a literati gloss, and perceived it as idolatry. Their iconoclastic rhetoric and actions were consequently more aggressive.

By examining the iconoclastic themes in missionary accounts, we may see how China was constructed in Western imaginations very much *as an idol*: materially fragile, lacking inner substance, and primarily concerned with surface display. Seen through the lens of Christian discourses on idolatry, Chinese religion was specifically constructed as meaningless performance, the "empty" worship of "hollow" idols. Such images of China in turn predisposed justification of Western imperial aggression, looting, and the spoliation of China's religious art. Acts of iconoclasm could also function as initiatory rituals to prove the sincerity of prospective converts, set alongside other measures of true conversion such as the repudiation of polygamy, the eating of meat, and the whole system of examination and probation.

While the judgement of idolatry was largely unilateral—Christians judging non-Christians—the practice of iconoclasm was not. A great deal of popular anti-Christian sentiment was channeled into material violence. It is often difficult to separate anti-Christian, anti-foreign, and even anti-Manchu sentiment, but whatever the reasoning, attacks on churches and clerics were frequent. Furthermore, specifically anti-Catholic discourses of idolatry and iconoclasm were disseminated by Protestant missionaries, with, for example, the result that the Taiping rebels were lenient on Chinese Protestants but harsh on Catholics.

The cumulative traditions of debate, as much as the Biblical sources, informed both Catholics and Protestants in their missions to China. Missionary journals make only very rare references to Byzantine iconoclasm, recalling instead the Bible, and in the case of Protestant missionaries, the Reformation.

There were, in fact, many Reformations, not all of which can be attributed to theologians. The targets of the protests were also multiple: indulgences, the use of holy water, the dominance of clerical religiosity, the power of the Church, mistranslation of the Bible, the use of Latin, the veneration of images, and belief in the efficacy of the Eucharist and the Mass. With its radical assertion of the inefficacy of any intermediaries or human effort to gain salvation, iconoclastic rhetoric in the Reformation was strongly dichotomizing. At different moments this either-or thinking drew sharp lines between God and the world, between the idolatrous and the non-idolatrous, between the Word and "tradition," or between Protestants and Catholics. The word "Catholic," in fact, became a "floating sign" which was applied far beyond the Roman Church, to "Indians, Jews, and Pagans" (Kibbey, 1986, p. 142), and even to certain Puritans (ibid., p. 145). The sign, inevitably, floated toward China. Western accusations of idolatry among the Chinese were inextricable from Protestant–Catholic polemics.

For some Reformation iconoclasts, representations *as such* became suspect. Calvin objected to the theater because it blurred reality and unreality; in this sense, bad acting was less harmful than good acting. The Mass in particular was a grand sham, whereas an austere, minimalist aesthetic in worship was a sign of purity. Calvin said: "wherever there is great ostentation in ceremonies, sincerity of heart is rare indeed" (*Institutes*, Book I, ch. 2, sect. 2). By the nineteenth century, China had been constructed in Western imaginations as a land of limitless ceremonial ostentation, and therefore insincerity and hollow illusion. James Hevia remarks, "To the reasoning eye, ceremony was contentless form designed to mask the mechanisms of power and awe the unenlightened mind" (Hevia, 1995, p. 74. See also Reinders, 2004a, ch. 7).

In the late Qing and after its collapse in 1911, increasing numbers of Chinese looked to the West, or to a partially "Westernized" Japan for their models of national and racial strength. Amid this pervasive mood of learning from the West, Christianity was often taken as a sign of modernity. Leaders of the new Republic of China, including Sun Yat-sen and Chiang Kai-shek, were Christian. Missionary journals expressed great joy at the improved receptivity of the Chinese to their message.

> China is awake and reforming herself . . . and so eager is she for schools that sometimes, almost ruthlessly, idols are expelled and nuns dismissed, and temple buildings utilized; so much so that many Buddhist priests have bound themselves by promise to establish modern schools in their courts, so as to preserve some hold on their property and homes. ("The Day of Opportunity and the C.M.S.," 1908, p. 132)

Missionaries were aware, however, that this openness might also allow in negative aspects of Western culture, such as access to "materialistic and Bolshevik thought" ("Encouragement and Opposition," 1925, p. 82), to "Lenin, Marx, Bertrand Russell, and others" ("The Influence of Students," 1925, p. 126), "infidel authors such as Ingersoll, Voltaire and Bradlaugh . . . Huxley, Spencer, Nietzche, Haeckel and Schopenhauer" (Mott, 1915, p. 114); "materialism added to materialism: progress, but no God!" (Blenkinsop, 1935, p. 127).

Vincent Goossaert has recently argued that that religious reforms proposed by Kang Youwei (1858–1927) in 1898 lay at the basis of large-scale religious destruction after 1904, and further during Kuomintang anti-superstition campaigns of 1926–7. "During the whole course of the twentieth century, the confiscation and destruction of temples was more than a side effect of social and political modernization; it was the effect of a religious policy, that is, a conscious, purposeful new relationship between the state and religious institutions" (Goossaert, 2006, p. 308). To explain this momentous transformation, which he claims marked the "beginning of the end of Chinese religion"—at least, in its traditional, premodern form—Goossaert argues, "Although openly confrontational in nature, Confucian fundamentalism and anticlericalism did not warrant wholesale temple destruction, whereas antisuperstition had such a destruction as one of its core projects" (ibid., p. 309). Certainly, anticlericalism and Confucian suspicion of popular religion have been staple fixtures of Chinese religious history. However, antisuperstition as defined by Goossaert was a new discourse—one that was deeply related to Western modern ideas about religion and its place in society.

Jesuit iconoclasm in China

The first major effort to evangelize China was undertaken by Catholics some decades after the beginning of the Protestant Reformation, and these missionaries were well aware of the Reformers' critiques of the Roman church. The best known of the Jesuits in China was Matteo Ricci (1552–1610), who gained the attention of the Chinese authorities not primarily by preaching the Christian gospel, but by expounding Western science and philosophy, and by his mastery of elite Chinese learning. His intention was to bridge the cultural differences through accommodation to elite culture, introduce Christianity as the culmination of Chinese culture, and thereby convert the Chinese from the official class downwards. He came to see in the Chinese Classics not only a religious, ritual, and ethical system of great beauty and subtlety, but also the prefiguration of Christ—God at work in a heathen culture.

This accommodationist view provoked the Rites Controversy, a debate over obeisance to ancestors and to Kongzi ("Confucius"). Ricci saw that obeisance to Kongzi was essential to the literati class, and indeed legally mandated for all government employees. In order to avoid the charge that the Jesuits were condoning idolatry, Ricci argued that these rituals were gestures of respect only, psychological rather than spiritual, secular rather than religious. The elite Chinese were not "worshipping" Kongzi. Finding support in some of the more secular-minded literati, Ricci claimed that bowing to the plaque (not image) of Kongzi was no more idolatrous than bowing to one's ruler, or to one's parents, or to a grave. Respect—both the emotion and the action—was in turn the fundamental building-block of the entire sociopolitical order. Thus, one effect of the Jesuit position was the attempted emptying of religious meaning from what otherwise would have been seen, fairly unambiguously, as idolatry: bowing to an object other than a representation of Christ, Mary, or the Saints. It may be that some of the persistent Western representation of Chinese culture as *surface* (superficial, empty, "face," excessively concerned with courtesy) begin with a Jesuit insistence on this crucial act as mere custom or form.[1]

While there is little evidence of the Jesuits themselves smashing images, they wrote in Chinese against idolatry, and undoubtedly incited Chinese converts to iconoclasm. The anti-Christian writer Xu Dashou condemned both popular worship of icons and popular Catholic iconoclasm:

> There is no end to the ignorance of the common people. If somebody comes and tells them that by flattering the hundred gods, they will obtain a hundred blessings, a heterodox cult immediately develops. If these people now come and tell them that by destroying the hundred gods [of the Chinese cults] and thereby flattering the Master of Heaven, they will acquire the greatest good fortune possible, the hundred gods will be abolished on the spot. (Gernet, 1985, p. 117)

An anti-Christian pamphlet complained:

> The most upsetting thing of all is that they destroy [the statues of] our Saints, cut off the heads of our deities, break the tablets of our ancestors and put a stop to [the continuance of] cults, endeavouring by all this to topple our Sages and Masters and sever our connections with our fathers and ancestors so as to sweep away all our moral principles and school traditions. (ibid., p. 129)

Elite Chinese culture already had a long tradition of hostility to the excesses of popular religion, to expenditure on Buddhist and Daoist temples, to wasteful sacrifices and unpredictable spirit possessions, and to popular cults as breeding grounds for rebellion. The Jesuit missionaries, themselves the products of elite education and clearly aligned with the upper classes in China, knew that their own rhetoric against idolatry would be perceived in terms of this native discourse against popular religion of the "uneducated masses" (Ricci, 1985, p. 439). In the *True Meaning of the Lord of Heaven* (*Tianzhu shiyi*), Ricci remarks, "I have read a great number of Confucian books and have noticed that they never cease to express animosity towards Buddhism and Taoism" (ibid., p. 101). Jesuit views of Chinese religions were strongly dichotomizing: Confucianism (with its residual traces of ancient monotheism, and finding its culmination in the Christian gospel) vs idolatry (Buddhism, Daoism, and anything else). Buddhism, however, warranted the greatest hostility.

Elite Chinese hostility to popular religion was probably less focused on the icon as such, whereas for Western Christians idolatry was a far more loaded term. "For the Jesuits, 'idolatry,' 'idol worship,' and 'sect of the idols' were all terms strictly referring to Buddhism" (Jensen, 1997, p. 47). Committed to the notion that the ancient Chinese worshipped the true God, Ricci and many others after him blamed Buddhism for the introduction of idolatry into China:

> Then, there were very evil men who employed witchcraft [*xiefa*] to subdue goblins and phantoms. Because of these unusual activities they called themselves Buddhas and Immortals and, relying on the dissemination of their precepts and arts, they deceitfully promised to bestow blessings and happiness in order to frighten and mislead the stupid and vulgar, causing them to make images of these men and to sacrifice to them. This was how it all began. (Ricci, 1985, p. 399. See also Reinders, 2004a, ch. 7)

Ricci was aware of native discourses and practices of iconoclasm, and noted the humiliation of images: "They do worship a few idols, but when these do not grant them what they desire, they beat them hard and make peace again with them later on" (Gernet, 1985, p. 82). On that topic, another missionary, Louis Le Compte (1655–1728), wrote:

> The Chinese are, however, sometimes weary of paying useless addresses to their idols, which are very numerous; for it often happens, that if after worshipping them a great while, the people do not obtain the blessing they desire, they use them in the most reproachful manner; some load them with hard names, and others with hard blows: we lodge you, say they, in a magnificent temple, we cover you with gold, and offer to you food and incense, and after all you are so ungrateful as to refuse our requests. They then tie the idol with cords, pluck it down, drag it along the streets, through all the mud and dunghills. Yet if they soon after obtain their desire, they then take the idol, and with great ceremony carry it back, and place it again on its niche, after they have washed it clean, fallen down before it, and made excuses for what they have done; promising that if it will forget what is past, they will gild it again. (Le Compte, 1787, pp. 104–5)

In other words, there was already the common practice of abusing an icon with the intention of provoking its power. Perhaps Ricci knew that any iconoclasm he or his fellow Jesuits might inspire would blend into these native practices, just as the Jesuits themselves had attempted to blend into Chinese categories (first as Buddhist monks, then as Confucian scholars). Indeed, as Gernet notes, "some Chinese, far from becoming upset at the ignominious treatment meted out to [the icons], reflected that it was, after all up to the gods to take their own revenge for such sacrileges" (Gernet, 1985, p. 84). Hence we find Christian-inspired iconoclasm accommodated to (or at least, perceived in terms of) a popular Chinese understanding of such violence. In later times, the inability of images to respond to abuse or violence became a recurrent motif in Protestant preaching. But Christian-inspired humiliation of images did not "make peace again with them later on."

Protestant iconoclasm in China

Jesuits in China had in their favor the power of persuasion, supported by some European science and intermittent imperial favor, but on the whole they were closely watched and tightly controlled. There was no strong military or diplomatic presence in China to back them up. Further stymied by the Papal decisions in the Rites Controversy, they succeeded in converting some Chinese, but not many. By the nineteenth century, the situation had changed considerably. Protestant missions in China developed in tandem with the increasing leverage of Western nations over the Qing state. The first Protestant preacher with the specific intention of missionizing China was Robert Morrison, sent by the London Missionary Society to Canton in 1807. As the century wore on, as trade increased and the Qing weakened, the numbers of missionaries bound for China increased exponentially, until numbers declined during the first half of the twentieth century.

Protestant missionaries invoked the same Biblical tradition against idolatry as Catholics, but with the added zeal of Reformation polemics. There are frequent references to the similarity of Chinese religion and Roman Catholicism. The hostility of Protestant missionaries to Catholicism took many forms, some of which were articulated in the earliest moments of the Reformation (attacks on images and ritual, etc.), and others which belong more properly to the missionary enterprise (the Catholic missionaries as competitors and obstacles to Protestant salvation).

Protestant missionaries worked within a transplanted sectarian rivalry that is clear from their accounts of meeting Catholic missionaries and Chinese converts to Catholicism. A typical comment makes this homology explicit: "We regret . . . that the Romanists, taking advantage of the close similarity which exists between their system and that of Buddhism, are endeavouring to transfer the natives *from one form of idolatry to another*" ("Romanism at Chusan," 1852, p. 11, emphasis added). A missionary, Frederick Foster Gough, visited a former Buddhist monastery that had been converted into a Roman Catholic Church, and remarked: "the gross idolatry of Buddhism had been supplanted—by what? by the alike idolatry of pictures, and the trust in false mediators" (ibid.).[2]

In fact, it was easier to convert those who had never heard of Jesus than to convert a Catholic to Protestantism. "The heathen, when perverted to Romanism, are much more inaccessible and indisposed to listen than they were before" (ibid. See also Lutz and Lutz, 1998, pp. 137–8). Missionaries described the puzzlement and hostility of Catholic Chinese upon hearing themselves accused of idolatry. Medhurst, speaking to a group of villagers near Shanghai, and

> finding that they were principally professors of the Roman-Catholic Religion, dwelt on the more prominent truths of the Incarnation and Atonement of Christ, to which they assented; but, on his subsequently enlarging on the necessity of trusting in Christ alone as the Saviour, and the sinfulness of raising other Saviours and Mediators—such as the Virgin Mary, who was only a sinful mortal like ourselves—they appeared to be somewhat staggered, and looked in his face as if incredulous and distrustful of his remarks. (1846, p. 90)

Protestant anti-Catholicism was transplanted to China and used to attack both Catholicism and non-Christian religious activity. Failing to acknowledge when Catholics or other "idolators" made the distinction between the sign and the signified, a good deal of Chinese religious practice would have seemed absurd or surreal to Victorian low-church Protestants. A mixture of sarcasm and horror pervades their anecdotes about icons, as, for example, in this story, by a concerned Protestant if not a missionary:

> China is the land of unreality. What would you think of people honouring a god whose head had fallen off? You would suppose that they would realize that that at least could not be the true god! But this is what really happened some years ago in the temple of the Great Mountain. The head of the largest idol there suddenly fell off, just as if his neck was broken! It was found that white ants had eaten through the posts that supported his great head! The body of the image was carefully carried to the back of the temple, and there buried, with the head that had fallen off, a mound being raised over his grave. But the trustees of the temple by no means lost their faith in a god who could not even protect himself from white ants. They sent out an appeal

> to religiously minded persons, and another large image was made to take the headless idol's place. A great many people came to watch the workmen put the new god into position, and a large sum of money was soon subscribed. (Foster, 1899, pp. 115–16; also Bettersby, 1897, p. 250)

The writer, Mrs Arnold Foster, seems incapable of understanding what is going on here, that the Chinese might not have perceived these events as their God's head falling off. Chinese religion is a topsy-turvy-land, full of pious fools and money-grubbing priests. We should recognize this kind of failure of hermeneutic goodwill not as a pernicious arrogance toward a semi-colonized and heathen culture (or not only as such), but as a systematic predisposition in the epistemological framework of missionaries in China.

The influence of aggressive Western nations behind the missions sometimes wavered, and some Westerners viewed missionaries as dangerous liabilities. In turn, many missionaries personally repudiated aspects of Western policy in China such as the opium trade and military adventurism. In general, however, nineteenth-century missionaries benefited from advantages such as rights of extraterritoriality, recourse to diplomatic channels, and outside funding. As the late Qing state grew weaker, missionary iconoclasm grew bolder.

Sometimes their enjoyment is clear: "It made a good smash, that white china idol! Destroying idols is worth-while work" (Pantin, 1925, p. 16). Protestants in particular sometimes made great public displays of iconoclasm. Some missionaries were happy to join crowds of converts in festive processions of destruction:

> Pa-Lau is the place where Mr. Wolfe was invited some years ago to perform "the interesting but novel ceremony of expelling the devil." A procession was formed, idols, incense-pots, sticks, &c., were collected and smashed upon a rock, and the fragments were thrown into a river close by, the waters of which speedily bore them away amid the apparent exultation of the villagers. (Stock, 1890, p. 211)

In other cases, missionaries were more isolated and cautious, but iconoclasm grew naturally out of preaching against idolatry. George Smith reported on a visit to a Buddhist monastery. As he remonstrated with the crowd,

> I proceeded to point to the idols with my umbrella; whereupon the principal idol gave way to the force with which, in my carelessness, I poked its various parts. The whole assemblage burst into a loud laugh, on which I was emboldened to show how little the other idols could help themselves. As I

> gave them a slight thrust they trembled, tottered, and tumbled from their thrones. The people again laughed heartily, as the priests tried for some time in vain to make one of the idols maintain its sitting posture, the fall having disordered its component parts. ("Buddhist Idolatry in China," 1847, p. 93)

This laughter of the crowd is difficult to interpret. People laugh for many reasons. It could be that the crowd was less "idolatrous" than missionaries thought, and found his disrespect amusing in a carnivalesque way. According to the missionary Medhurst, when they are told that idolatry is absurd, the crowd just laughs (Medhurst, 1846, p. 91). In another case, missionaries visit a temple, and in the crowd are some priests: "I asked one of them whether the rites they were performing were true or false. He made no answer, but a brother priest shouted out, 'False,' and laughed" ("Missionary Labours at Ningpo, China," 1859, p. 31). In another case, the crowd laughs to see a colporteur working for the missionary Wolfe put sweetmeats into the mouth of an icon, to see if it could actually eat them ("Ku-cheng," 1868, p. 3. See Figure 3.1).

Figure 3.1 "Feeding the Idol." From *Quarterly Token for Juvenile Subscribers*, 48 (January 1868), p. 3.

Mockery or abuse of icons may not have been shocking blasphemy, but rather titillating transgression, or just free entertainment. Missionaries were sometimes justifiably nervous about their provocative actions and teachings. After a very destructive anti-Christian riot in 1869,

> It was pleaded by the mandarins that Siek [a Chinese convert] and two or three others had wantonly destroyed the idols in one of the temples; but this was never proved, though it seems probable that some of the converts had been more zealous than discreet in their conduct. (Stock, 1890, p. 108)

Iconoclasm by Chinese Christian converts

Though their rhetoric was relentlessly iconoclastic, cases of foreign missionaries themselves destroying Chinese religious images were far outnumbered by the destruction of images and ancestral tablets by Christian converts. Missionary Mary Darley made this explicit after an act of iconoclasm: "That was the only time I destroyed an idol myself; we missionaries invariably leave this to the people themselves; it is far better that it should be so" (In "Our Twenty-Sixth Anniversary Meeting," 1906, p. 86). Most conversion anecdotes include the note that they "gave up their idols," without further elaboration. For example: "Their 'kitchen god' was burned, and many idolatrous practices given up" (Main, 1887, p. 3. See also Stock, 1890, pp. 88, 98, 99). On the one hand, the repudiation of all indigenous deities could be very systematic:

> On the 22nd, they abjured the idols as well as the god of the hearth and the god of the swine in the presence of Jiang's brothers. In the presence of Jiang's mother, they abjured the god of the hearth, the earth god, the dragon spirit, and the doors gods. In the presence of Jiang Dade they renounced the spirits of the heavens, the hearth, the dragon and the doors. (Lutz and Lutz, 1998, p. 18)

It was not always enough simply to shrug off "idolatry." In some cases, each god required a specific renunciation, suggesting a demonic rather than materialistic model of idolatry. Possibly too the need to abjure each of the many gods suggests that "idolatry" in the minds of the converts was not a categorical rejection of everything but the Christian God, but rather a need to sever a network of relationships—rejecting the hearth god did not necessarily imply a rejection of the door gods. Just as conversion meant severing social obligations such as contributing to the general fund for festivals, becoming a Christian meant ending long-term relationships with the spirits. Note also that this testimonial carefully records the witnesses, as the acts of repudiation had juridical qualities. The destruction of idols seems to have been publicly known. Even while a convert's fellow villagers readied for a feast to honor the gods,

> Many admitted that it was useless to worship the gods. On the other hand, they were absolutely unwilling to part with the veneration of the ancestors. Trusting in the protection of God, we tore down the names of the false gods, or the spirits of the hearth, of the earth, of the skies, of the shrines, and of the dragons in the homes of my mother, older brother and several of our relatives. We also buried the bones of my father and his brother which, according to the superstitious custom of the Chinese, must be exhumed in order to be preserved in earthen vessels, after which they are reburied. We did this without any kind of heathen ceremony and without attention to a special time and place. All this outraged the villagers, who were certain that this turn of events would bring great misfortune to them. (Lutz and Lutz, 1998, p. 17)

Sometimes these displays were less openly public, and took place in the missionary compound or "on the topmost of the three stone steps leading into the [Worship] Hall" ("Our Twenty-Sixth Anniversary Meeting," 1906, p. 86). Some converts were apparently afraid to destroy their images at home. D. A. Callum recounts one such farewell ceremony:

> On Sunday, July 7 [1907], our boys' school teacher brought his idols to be burned. He had been an enrolled hearer for over a year, and had given up all faith in idols, but had not had the courage to take down all the family idols, as some members of his family were opposed . . . The next week the "tablet to Heaven and Earth" and a number of idols were brought to the church, and were burned before Morning Service on July 7. We all stood around and sang "Jesus shall reign" while the idols burned. (Callum, 1907, p. 188)[3]

On most occasions, the images destroyed by Chinese converts belonged to their own household, but there are also cases where converts attacked the images of others. Wolfe recounted a story of a convert called Old Chung-Te during a fire:

> As the fire drew nearer and nearer, the neighbours brought forth all their idols and placed them in rows before the flames to stop their advance. Some of these idols were placed in front of Chung-Te's house. This aroused the old Christian's zeal, and in the presence of the astonished multitude he took the heavy mattock with which he wrought in the fields, and with it belaboured

> the bodies of the idols, and in a short time demolished them into broken pieces, and raising his hands to heaven, he called upon "the great Creator, the true God, his Heavenly Father," to save him and his neighbours from the approaching flames. ("The Fuh-Kien Mission," 1884, p. 105)

The wind changed. He and his neighbors were spared. Another convert, ca. 1866, "brought out his household gods and ancestral tablets, and publicly burnt them in the presence of the horror-stricken villagers. Recovering from their surprise, they rushed upon him, but he escaped" (Stock, 1890, p. 147).[4] This public act was followed by a remarkable vision of divine iconoclasm:

> He dreamed that he was in deadly conflict with the devil, who was pressing him very closely and gaining the victory. Suddenly, as he was about to give up in despair, he heard in the distance the shouts of an immense army,

Figure 3.2 "Chinese Converts Bringing Their Discarded Idols to a Missionary." From *Church Missionary Gleaner* (September 1, 1900), p. 132.

> increasing in terribleness as it appeared to approach him. He took courage, for he perceived that the devil evidently got alarmed and relaxed his grasp; and as the mighty host came nearer, the enemy fled in dismay. The army approached, shouting and exulting, and passed on to the village temple and tore it to dust and destroyed all the idols. The leader of the army came near, smiled and said, "Be of good courage; fear not, but believe," and passed away. (Stock, 1890, p. 147)

The devil fled unharmed, but the army of God destroyed the devil's palace and material representations, reminiscent of the destruction of the Yuanming Yuan a few years before.

The fear of illness or death as punishment for iconoclasm is often mentioned, as well as the consequent bravado of iconoclasts. For example, the preacher Zhang Fuxing (1811–80) found a local festival taking place in his hometown. In violation of protocol, he ate some of the sacrificial meat, and then condemned the idolatry. He said: "If Tai Zi has power, then let him punish Zhang Fuxing. And if, on the way home, Zhang Fuxing starts gushing blood or develops stomach cramps, he will then believe that Tai Zi has power. Otherwise, Zhang Fuxing will believe in his own words" (Lutz & Lutz, 1998, p. 37). "He then returned home to destroy his family's idols, including that of the much revered Guan Yin" (ibid.).[5]

Fear of idols, once overcome, could turn to vindictive indignation. Mary Darley wrote:

> I placed the idol on the topmost of the three stone steps leading into the Hall, and asked one woman to go home and bring back her sharpest wood hatchet. Frightened, she brought her hatchet and handed it to us. Then these women found voices, and they shouted, "O sisters, you are not going to hit it! If you touch it it will cry out; it has great power, it has done many things." So we said to them, "Bend near the idol," and they bent down. Then I said, "When it gives its first cry we will stop hitting it, but until it does do so we will go on." Perhaps our strikes were not very heavy, or else the wood of the idol was very hard, so that we spent some little time hitting it without any effect, but at last the head came in two. Then there was a shriek from the woman "It has not cried! Verily, what the sister said is true. Hit it! Smash it! Do not leave one bit remaining!" ("Our Twenty-Sixth Anniversary Meeting," 1906, p. 86)

Missionaries often regarded the destruction of images as largely favorable to one's health. Indeed, idolatry was regarded as physically unhealthy and iconoclasm could be an agent of healing. In general, Victorians were quite prone to merge the imageries of disease, idolatry, and immorality. There was also a common blurring of idol temples and opium dens under the common image of the "den of iniquity" (Reinders, 2004a and 2010, ch. 3). In one anecdote, the mother of a "witch-doctor" is possessed by a demon; a Christian relative tells him Jesus can cast out all demons. He prays and the demon leaves, but there is concern that it might come back. The Christian points out all the gods in the house. "You will have to get all these gods removed" (Goforth, 1931, p. 104). "You get rid of all these proofs of devil worship, and you and your whole house decide to serve the Living God, and the demon will not have a chance to come back here where the Lord reigns." They agree: "They tore all the gods down and smashed the incense-pots and threw them out" (ibid.). Such stories of demonic possession must be situated in a larger Victorian discourse on Christianity as healthy, vigorous, and masculine.

Anti-Christian iconoclasm

While missionaries directed their attacks at images, they were also "preoccupied with social behavior" (Maurer, 1951, p. 127), and criticized many other social phenomena such as foot-binding, opium, and the recalcitrance of officials. But the core of their entire mission, as they understood it, was to save Chinese souls. While foot-binding was condemned for many reasons, salvation did not depend on feet. In contrast, idolatry was a sin that prevented salvation, and hence, Protestant iconoclasm was focused on religious images. While they facilitated the burning of household icons, and cleared away images to make room for Christian worship, I have never found any account of missionaries destroying entire buildings or killing people. In contrast, very little anti-Christian iconoclasm specifically focused on Christian religious objects; it seems more often to have been directed against buildings, furniture, books (including the Bible, obviously), missionaries themselves, and Chinese converts. For example, describing events in 1869, Stock describes how "a body of the Chinese soldiers and police, accompanied by some of the gentry and *literati,* attacked and broke open the Mission chapel, destroyed the furniture, and seriously damaged the building" (Stock, 1890, p. 108). They went on to plunder the homes of Chinese converts. The destruction of buildings was often accompanied by looting, whether of the church compound or of converts' houses. Missions were attacked because they were the most visible local manifestations of a resented foreign presence in China.

There are many accounts of rioting mobs destroying churches and mission compounds. (For example, "Outrage on the Mission at Fuh-Chow," 1878, pp. 137–8; Thompson, 1900, pp. 148–50.) As one would expect of a basically disorganized mass movement, anti-Christian action tended to operate in multiple ways. In one report, we can see the variety of hostile actions. In 1892, what Rigg calls "filth" (presumably feces) was dumped and smeared on his door. Stones were thrown at the hospital, and as Rigg tries to escape he is mugged, knocked down repeatedly, and thrown him into a "pon k'ang" or "filthy pit." At the mission hospital, "everything was smashed, broken, and burnt." The houses of various Chinese associated with the mission were "injured" and undoubtedly plundered. "They smashed all the contents of the hospital to atoms, and burnt all garments and bedding" (Rigg, 1892, p. 138).[6] At least in Rigg's telling of the story, none of this destruction targeted Christian symbolic objects such as crucifixes.

The destruction of Christian books took place constantly, and not usually during riots. "Of the thousands of books and tracts distributed, few have ever been heard from" (Bush, 1865, p. 97). What happened to these books? Some were put to practical or profane use: "making soles for cloth shoes, employing them at the toilet, papering walls, and so forth" (Lutz and Lutz, 1998, p. 74). But the destruction of so much pious literature so intrigued Matthew Tyson Yates that he became a kind of detective: he first distributed tracts along a street; a month later there was no trace of them. Some said they had passed them on to friends to read, but Yates did not believe this. Then he got a tip-off from a Chinese friend to go to a certain temple. Hiding himself away, very early one morning, he saw seven or eight coolies bring in sacks of books, including his own. "These loads of books were to be burned before the idol, and some of the ashes distributed on the waters of the canals and rivers, to furnish the spirits of the departed with reading matter, and the balance, mixed with oil, would be used to make the paste of which the smooth surfaces of sign boards and lacquered ware are made" (In Yates et al., 1879, p. 112). The spectacle of the Bible burnt as an offering to Buddha and distributed for watery souls to pass the time is wonderfully ironic.

More tragic, of course, was the destruction of lives during anti-Christian and anti-foreign violence. Again, the literature is voluminous on killed missionaries and (much less so) killed Chinese converts. While missionary organizations did not want to deter their readers from undertaking foreign missions, the deaths of missionaries were undoubtedly good for raising funds and rallying readers to the cause. However, public sympathy was never unanimous, and British hostility to missions was fueled by such deaths, especially when they led to further military expenditure and disruption of trade. Some felt missions were a dangerous annoyance or at least a waste of money. It was therefore very important to refigure murdered missionary as martyrs. An 1895 Resolution of the Church Missionary Society General Committee included rhetoric of "the blood of the martyrs" becoming "the seed of the church" ("Resolutions," 1895, p. 142). After the Ku-cheng martyrdoms of 1895, anti-Christian violence was interpreted in Biblical and apocalyptic terms: noting a banner proclaiming "The dragon makes war with the foreigners' saviour," Miss Tolley commented, "This is fulfilling the words of the Bible that the dragon shall make war with the Lamb" ("Foochow—Extracts from recent Letters," 1895, p. 474. She is referring to *Revelation* ch. 12). The *Church Missionary Gleaner* (later, *Outlook*) and other publications gave space over to various forms of commemoration, such as pictures of the martyrs' graves ("Scene of the Ku-Cheng Massacre," 1895, p. 165).

In 1864, during "a violent outbreak of popular fury" (Stock, 1890, p. 142) against missionaries, "The rioters destroyed a preaching chapel, schools, Mission-library, and dwellings of the Native agents, did much damage to other property, and inflicted severe injuries on such Chinese Christians as they could lay hold of" (ibid., p. 33). Then, however, Stock supplies the higher rationalization of this violence: "the riots did a real service to the work by bringing Christianity prominently before people of all classes . . . Crowds flocked to the rebuilt chapels" (ibid.). Similarly, "The vine, pruned by the sharp knife of persecution, and with the unfruitful and withered branches cut away by excommunication, shot forth its boughs and yielded its grapes plentifully" (ibid., p. 50).[7]

Taiping iconoclasm

An interesting example of the mixed messages of native Chinese and Christian iconoclasms was the Taiping Rebellion (1850–64). The Christian influence on the ideology of the Taipings was in most respects shallow. Wehrle remarks on its "pseudo-Christian origins" (1966, p. 7) and Boardman (1952, p. 79) notes the tendency to use Christian symbols out of context. The spiritual head of the Taipings, Hong Xiuquan, believed in Jesus, but also believed himself to be Jesus' younger brother. When the Taiping theology became widely known from 1853 on, missionaries on the scene shared this evaluation of Taiping theology. Some Protestant missionaries, however, expressed enthusiasm for the rebellion, specifically their program of anti-idolatry: "they seem rather struggling for religious liberty, and are *really upsetting idolatry*! I now begin to sympathise with them in their struggle, and to look for important results" (I. J. Roberts, in

Clarke and Gregory, 1982, p. 20). In an 1899 book, Mrs Arnold Foster recounted the Taiping story with approval of its iconoclasm. Little mention is made of the bloodshed caused by the Taipings, and violence around Shanghai is blamed not on the Taipings but on a Triad group (Foster, 1899, p. 132). She is almost nostalgic about the rebellion, and regretful that it failed. If they had succeeded, "China would probably have been a very different country to-day from what it is. But the long-continued fighting and the unsettled state of the country were bad for trade" (ibid.). She seems to be blaming mere commercial interests for ending "that remarkable movement, which might have led, had it been successful, to the abolition of idolatry throughout the empire" (ibid., p. 133). "But the power of the Tai Pings was crushed by the help of Christian soldiers from the West, and the Peking Government is as heathen to-day as it ever was" (ibid., p. 134).

Foster admits that Hong's theology was a dangerous mess, and that millions of people died. "Still, the results of the movement were not all evil. Idolatry received a severe blow in the nine provinces through which the Tai Ping army passed, and all along the valley of the Yangtse, wherever we go now, we find ruins of temples and shrines that have not yet been rebuilt, and priests will tell us that things have never been the same since the coming of the long-haired rebels" (ibid., pp. 134–5).

Griffith John wrote with approval, "They hate idolatry with perfect hatred" (Clarke and Gregory, 1982, p. 272). The Taipings were seen obscurely as "a miracle of what the dissemination of the Scriptures unaided by instruction had performed" (Boardman, 1952, p. 4). The evidence of this miracle was primarily their attacks on Buddhist and Daoist temples, as well as their suppression of brothels, gambling, and alcohol. The missionaries of Ningbo withdrew from the violence but remained "on the watch for opportunities of being useful to these Taepings" ("Tidings from Ningpo," 1862, p. 44). The Manchu government and other observers naturally considered the Taipings to be basically Christian, in part because of the use of Christian terms, but more importantly because of their iconoclasm:

> The government of the celestial empire is beginning to believe that the leaders and instigators of the rebellion are Christians, and supports this suspicion on the fact that, of all the religions in China, the Gospel is the only one professing the hatred of idols and the worship of them. (Rizzolati, in Clarke and Gregory 1982, pp. 30–1)

Taiping troops called missionaries "foreign brothers" (*yang xiongdi*), and some missionaries took the connections to heart: "Protestant Missionaries in China!" pronounced Griffith John, "This Insurrection is your offspring. From the want of your parental care, it has grown deformed, and wayward; it still possesses the elements of a perfect man" (Clarke and Gregory, 1982, p. 278). However, relations between Taiping leader Hong Xiuquan and the few missionaries who met him were frosty at best. Eventually, the Taipings were commonly regarded as "desperadoes of the lowest class" ("Passing Events in China," 1854, p. 2) or "bloody iconoclasts and merciless vandals" (Woodbridge, 1919, p. 29). But it is no surprise, then, that during the period of Taiping iconoclasm, all Christians in China were to some extent implicated as rebels, because they were similarly iconoclastic (Lutz and Lutz, 1998, p. 40).

As we have found in most other cases of iconoclasm, the destruction itself often made a mockery of religion:

> Wherever they come, they overthrow and destroy to their very foundations the temples of the idols; the break up, trample under foot, and reduce to dust the so-much venerated gods of the people. The monasteries of the Bonzes and Bonzesses are no less summarily dealt with. After having sacked and demolished their convents, the insurgents parade their divinities in a sort of masquerade, and make a complete carnival of their idols and other objects of their superstition. (Rizzolati, in Clarke and Gregory, 1982, p. 30)

And also: "The rebels have shown their scorn of the idols, by chopping off their noses, and placing them in ridiculous attitudes" (Edkins, in Clarke and Gregory, 1982, p. 282). After watching 50 young men pull down a temple, Edkins notes, "Truly these young rebels have their amusements, a pleasant change from the war of Cannon and the brunt of battle" (ibid., p. 285).

The "playful" or theatrical qualities of some acts of destruction are clear. Hong and some of his early followers attacked King Gan, a local deity. "Shouting abuse at the bearded image of King Gan in its dragon robe, and striking at it with his staff, Hong lists ten counts of immorality of which the idol has been guilty" (Spence, 1996, p. 100). Then,

> with his willing helpers, Hong topples the image from its resting place, stomping its hat, pulling out its beard, shredding its dragon robe, digging out the eyes, and breaking off its arms. Festooning the desecrated shrine with their triumphant poems, and posting on the wall a manifesto of defiance to the devil demon Gan. (ibid.)

To justify their actions, the Taiping leaders used some of the same sources as earlier Christian iconoclasts, by way of Christian tracts and translations of the Bible. For example, Hong Xiuquan drew from Psalm 115:

Their idols are silver and gold,
the work of men's hands.
They have mouths, but do not speak;
eyes, but do not see.
They have ears, but do not hear;
noses, but do not smell.
They have hands, but do not feel;
feet, but do not walk;
and they do not make a sound in their throat.
Those who make them are like them;
so are all who trust in them.[8]

Taiping attacks on iconicized bodies of monks and nuns tended to deface rather than obliterate, attacking the monastic *habitus*. Sources seem unanimous on the rebel requirement that monks cease to shave their heads. Captain Fishburne and company "found the priests there, and they stated that they had not been injured; they were given books, and informed that they must allow their hair to grow—their practice being to shave their heads" (Clarke and Gregory, 1982, p. 57). Mrs Arnold Foster quotes a Mr Meadows, interpreter on board the *Hermes*. "Further, the few scared priests who followed my steps had the hair growing all over their heads, and told me that the rebels had prohibited them on pain of death from practicing the monastic rite of shaving" (Foster, 1899, p. 130–1).

Regarding one of China's strongest iconoclastic tradition—elite Confucianism—Hong had qualified praise for its destruction of temples. He said: "One cannot say these people did not have acute awareness; but what they destroyed, burned or criticized was limited to certain lascivious shrines, Buddhist practices, and improper sacrifices, so that everything they did not destroy, burn or criticize remains with us to this day" (Spence, 1996, p. 97). Nonetheless, Hong's early group began with the images and religious sites most immediately familiar to them, the tablets to Confucius and his disciples in the local school (ibid., p. 68). They argued that there were two causes of the Chinese lapse from the primordial monotheistic worship of Shangdi: "a clouding of the former clarity of Chinese perception, and a deliberate deception on the part of Taoists, Buddhists, and the Devil himself" (Boardman, 1952, p. 67). The historical support for this argument reads like standard neo-Confucian anti-Buddhist history. Famous anti-Buddhists were extolled: Emperor Wu of the Northern Zhou (r. 561–78), and Han Yu (786–824), for example (ibid., p. 68).

Hong's early activities were primarily directed against idolatry rather than against the Qing. As the scope of the rebellion increased and warfare engulfed China, the Taiping armies continued their massive program of iconoclasm. The Qing state was itself pulled into the iconoclasm, confiscating Buddhist metal (ritual objects and images) to mint coins (Spence, 1996, p. 249). In the Taiping and other rebellions in the late Qing, military action of any kind often led to the destruction of temples and churches—along with houses and shops. During the Boxer rebellion of 1900, Western armies destroyed temples in Beijing, not primarily with religious motives but because the Boxers often used Daoist temples for refuge. Nigel Oliphant describes one such attack.

> we forced our way in and found about forty Boxers in the inner building of a small Taoist temple. We poured bullets into this freely, and finally broke down the wooden front and door, and finished off what few Boxers remained. It was absolutely a slaughter, but we found the mangled bodies of a few prisoners, and therefore knew no mercy. In the end there were forty-six dead bodies in the temple, and we then went marching all through the East city, passing all the burnt mission-houses on our way. (1901, p. 16)

Later that day, Oliphant returned to the temple, "but did not stay long, as the place was beginning to smell. It was a ghastly sight, those forty-six bodies lying huddled together inside the temple with the hideous Taoist gods grinning above them" (ibid., p. 18–19, also p. 56). Other acts of iconoclasm during the siege of Beijing were specifically aimed at the metal: they scrounged for metal to use for bullets and balls: "All the temple candlesticks, vases, images, in fact everything that can be melted, have been gathered and molded into ammunition" (Conger, 1909, p. 121). Some of the destruction during the Siege was preemptive. Oliphant's journal notes: "There was nothing much on in the morning except that I set fire to the Red Temple in the Hanlin for Poole, and a very successful fire it was" (1901, p. 55).

China in decay and the necessity of violence

Even the massive violence of the Taiping Rebellion was but a temporary stimulation to the more natural state of inertia:

> A mighty movement has taken place amongst the Chinese themselves. That there should be a rebellion in China is nothing wonderful. Rebellions in

> China . . . have been ordinary occurrences. They seemed to be short-lived disturbances of the inert mass, which soon subsided, and all became stagnant and monotonous as before. ("The Movement in China," 1853, p. 110)

This was a common view. Whereas Western violence could fundamentally change China, China's own "indigenous" violence did not have that capacity. The West was regarded as the only source of progress, while China inevitably tended to entropy.

Anti-Catholic traditions, so easily transferred to Chinese religions, further legitimated the prevailing Victorian representations of China. If the Jesuits and the fad of Chinoiserie had put China on a pedestal, Victorian writers succeeded in knocking it off. One frequently finds the most lurid descriptions of China:

> In every place we came to the buildings were in an extensive state of dilapidation, and a Chinese city looks to the eyes of a Western barbarian like an immense mass of ruins, covered with an unbounded population wallowing in filth and thoroughly enjoying it. (In Stock, 1890, p. 60–1)[9]

A Chinese city is "a labyrinth of tottering decay such as Dickens would have loved to portray" (Couche, 1915, p. 185). Also: "China is a stagnant pool. It needs only to stir it, to be convinced how foul it is" ("Chinese Justice," 1853, p. 33). The image of China as ruined antiquity had a strong precedent in early nineteenth-century images of Greece and the "general atmosphere of pillage" of Greek antiquities (Bracken, 1975, p. 39). The Catholic traveler Evariste Huc noted the Putuo temples were in disrepair: "some [pagodas] are fallen entirely into ruins, and the gods lie prostrate, with their faces to the ground, and serve sometimes for seats to the curious travelers who visit this holy site" (Huc, 1855, p. 197). China's gods lay moldering between the dirt and Westerners' rear ends. Visiting the same island, Medhurst remarked,

> On all sides, I was gratified with perceiving marks of decay in the temples and adjacent buildings, and earnestly hope that future travelers will find these worse than useless structures level with the ground, and the lazy drones who inhabit them scattered among the useful and intelligent part of their fellow-men. (1836, p. 411)

In 1937, a missionary visitor noted "there, in a coffin dump, and amid a lot of washing hanging out to dry, were the idols that had been cast out of the city! They were in all stages of decrepitude—it was a pitiable sight to see an idol without a head, not at all like a human form, and with an incense stick still placed in front of it!" (Barclay, 1937, p. 175).

Throughout the nineteenth century the prevailing rhetoric construed China as place of decay and ruin, stagnation and rot. Not only its buildings, but according to E. P. Thwing, the "whole fabric of heathenism was honeycombed and some day would disintegrate" (Lewis et al., 1890, p. 31). In this view, the ancestral cult, the government, Confucianism, and all manner of superstitions conspired to keep the Chinese contained within strict boundaries, resistant to the globalizing forces of trade and the spread of European civilization, and this imprisonment led to an unnatural stagnation of the entire culture. Travelers remarked on the ruined state of many pagodas and temples; these images of decay merged with representations of Chinese bodies as decaying: the meticulous attention to graveyards (Fortune, 1853, vol. 1, pp. 78–9), foot-binding, beggars, opium-addicts ("Revolting Scenes in the Beggars' Square, Canton," 1845, p. 137), charnel pits, and even descriptions of heaps of hacked and mangled corpses ("China—Distress of Nations with Perplexity," 1862, pp. 78–9). These bodily images blended with a prevailing judgement about the rotten condition of Chinese culture, religion, and political order. This rot, it was felt, could not be reformed so easily, and there were often remarks about the need to "shake up" China, to give it some blow or slap, to force open its doors so that the "fresh air" of Christian manliness and trade could enter. Missionaries and travel writers attributed to the bureaucracy a belligerent resistance to any such penetration, so that violence was necessary for their own good. Punning on the name of the Society for the Diffusion of Useful Knowledge, S. Wells Williams wrote in 1858: "I am afraid that nothing short of the Society for the Diffusion of Cannon Balls will give them the useful knowledge they now require to realize their own helplessness" (1907, p. 257). Gleefully contemplating the Taiping rebellion, the American Baptist missionary Rev. I. J. Roberts argued that "nothing less than a mighty shaking and rending" would let China find the Gospel (Clarke and Gregory, 1982, p. 21).

Although by no means universal, images of violent penetration pervade the rhetoric of evangelism in China. Goodrich comments, "When we go to preach in our chapels, we want at least one *great thought bullet,* rammed down with argument and illustration, and, behind all, the power of feeling, the power of the Holy Ghost, to carry the thought straight into the hearts of men" (Yates et al., 1879, p. 87).[10] Truth was a bullet in the heart! Another missionary, Samuel Dodd, at the same conference remarked that the Second Coming would "give heathenism its death blow" (ibid., p. 109). Another, S. L. Baldwin, said, "good guns are true civilizers when aimed at the towers of despotism in a righteous cause" (ibid., p. 209). In an essay titled "The Accelerated Momentum of Truth,"

E. P. Thwing, wrote of the inevitable progress of truth and the folly of trying to stop it, imagining progress as violent against human bodies: a peasant standing in front of a train and getting run over; a man putting his foot out to stop a cannonball and losing his leg (Lewis et al., 1890, p. 29).

The idea of the necessity of benevolent violence was influenced by the Victorian emphasis on parental discipline: "spare the rod and spoil the child." Furthermore, the ethos of these missionaries was strongly influenced by the Christian Manliness (or "Muscular Christianity") movement (see Hall, 1994; Houghton, 1957, pp. 201–9). These terms refer to a broad-based cultural movement that associated physical strength with spiritual strength, vigorous activity with religious certainty, masculinity with Protestantism. In this view, championed by Charles Kingsley (1819–75), the Reformation was moment of union between true religion and "Primeval Teutonic energy" (Wee, 1994, p. 74). Hence we find missionaries in China voicing this masculinist religiosity: "Christianity is essentially aggressive and revolutionary" (Yates et al., 1879, p. 388). "The kingdom of Christ can be set up only by aggressive work. Missionary work is aggressive work" (ibid., p. 339).[11] According to widespread colonialist assumptions, the inherently vigorous Northern Europeans had a duty to bring life to the lazy, torpid cultures, a "white man's burden" justified by Nature: "By a law of nature when a thing becomes stagnant it soon grows corrupt. The same law holds good among men" (ibid., p. 340. See also Houghton, 1957, p. 243–6).

The violent penetration was inevitably gendered as the taming of a haughty female. Reflecting on changes since 1877, Young John Allen noted in 1890 that China had been humbled: "her exclusiveness was penetrated; her isolation uncovered; her supercilious bearing rebuked; the high prerogatives she had assumed were abased" (Lewis et al., 1890, p. 12). In the comments which depict China as a haughty, supercilious woman who (for her own good) is "abased," "rebuked," "uncovered," and "penetrated," one can also sense a class consciousness, mixed with a sadistic eroticism. Even if the suggestion of rape mentality is rarely so explicit, missionaries consistently gendered idolatry and obeisance as female, just as they feminized all of Chinese culture: "it is [the non-Christian priest's] policy to keep his followers in a gross and credulous state, the women especially, who, in all countries, are easy captives to superstition" ("Sketches of Idol-worship," 1874, p. 117). Many other examples of this kind of remark could be cited.[12] One finds repeated the basic assumption that idolatry is distinctively female, and therefore especially degrading if practiced by men, who ought to have a greater propensity for rational thought.

Continued influence of Christian-inspired iconoclasm

Missionaries often complained about difficulties in renting or buying places to live or preach. In some cases, they were able to build houses and churches, but in the early days they had to take over existing facilities. Among these buildings were Buddhist and other temples. In one such case, missionaries bought an old Buddhist temple to use as a church and initially worshiped in it "surrounded by the old trappings of heathenism" (Hoare, 1894, p. 91), though the idols themselves had been removed, perhaps (as in another anecdote) "shut up in cupboards at the rear of the building" (Jackson, 1912, p. 158). Or in 1907, C. Shaw describes looking for a place for Christian worship, and the only available place was "full of idols and ancestral tablets"—which they promptly cleared out (Shaw, 1907, p. 42). Apparently this kind of take-over was not always complete. Stock describes a village where the ancestral hall had been divided in two: "The Christians have one side; the heathen the other" (Stock, 1890, p. 132).

The reforms of the late Qing, and after 1911, called for the establishment of a widespread system of new schools, and in many cases Buddhist temples were used for this purpose. Missionaries enjoyed what they saw as a victory of progress and Christian civilization against the forces of darkness, and were inclined to take some of the credit. O. M. Jackson describes how "This great system of Heathenism" was weakened by the Chinese reforms, and "by the direct impact of the work of Christian Missions" (Jackson, 1906, p. 108; see also "Schools in Temples—Idols Hidden by Blackboards," 1907, p. 188–9). He remarks—perhaps "gloats" is not too strong a word—how new elementary schools were paid for by taxing temples, and by using temple buildings. The changing policy made some missionaries view the new national or provincial governments as allies. "With regard to idolatry, I think it is not too much to say that it has received its death-blow nationally, though individually and locally there is apparently as much as ever" (Boaz, 1912, p. 216). The provincial administration of Yuan Shikai was particularly active in converting temples to schools, as part of modernizing campaigns in collaboration with rural elites (Duara, 1995, p. 96). "Instead of burning the temples, as the Taipings did, the revolutionists converted these dark, germ-breeding buildings into primary schools, after making windows in the walls" (Woodbridge, 1919, p. 110). Buddhist efforts to create a Buddhist modern education system, such as the work of Tai Xu, did not go very far.

The "modernity" so eagerly pursued in the first decades of the century was also marked by scientism—a faith in the power of "science"—and by the common post-Enlightenment model of science and religion as necessarily antagonistic. Certainly among the most vocal of educated Chinese, this was an era of unprecedented hostility or at least disdain for traditional Chinese religions. In addition to the anti-Confucianism of the 1920s and 1930s, there were many organizations that aimed at a much larger target. The May Fourth movement rejected traditional literary models and is often labeled iconoclastic, though in the more abstract or metaphoric senses. In that tradition, and as part of a power struggle against Chiang Kai-shek, South China in the period 1927–30 saw an iconoclastic campaign "spearheaded by the left wing of the nationalist Party" (Duara, 1995, p. 98). For example, Mrs Gear Willett notes that "The old temples are being pulled down in many places—not everywhere—and it is quite usual to see 'gods' in groups, left exposed to the weather, after the destruction of some place of worship" ("Present Conditions in China," 1931, p. 73). The rhetoric of this campaign used various versions of an evolutionary paradigm, with a strong sense of a break between premodern and modern. "Just as the monarchy was unsuited to the present age, so too was religion unnecessary in an age in which science had proved the nonexistence of God" (Duara, 1995, p. 100). The Social Darwinism of this movement later mingled with Marxism. Organizations such as the Anti-Religious Federation and the Great Federation of Non-Religionists were formed (ibid., p. 99). The traditional lunar calendar was banned in 1912 and replaced by the Western (Gregorian) one. The rapid emergence of women's literature and discourses of women's liberation, and the increased priority on young people—these implicitly or explicitly represented repudiation of the Confucian past. As a 1930 article in the *Church Missionary Outlook* commented,

> In every department of life, religious, social, political, what cannot stand the test of reason is condemned and abolished. This is why temples are being destroyed or transformed into schools and museums, idols are being beheaded and burned, fortune telling, geomancy, necromancy, are declared illegal; in short, all old customs and traditions are in the melting pot. (Norton, 1930, p. 70)

The same year, Mary Darley reported:

> A few days ago this city thrilled with apprehension when it was known that the "local Government Board" had sent soldiers to destroy all the idols—over one hundred—in the "City King god's temple." We went to see the scene of desolation. Idols great and small, whole and in pieces, lying about in the various halls. The King god himself prostrate, half of his face hacked off, his red, embroidered satin robe on the ground near by. It was indeed the

> moment in which to point the awed and wondering onlookers from the "dead idols" around them, to the "Living Creator God." (1930, p. 47)

Nonetheless, anti-Chinese-religion turned easily into generalized anti-religion. Chinese "students are exposed to anti-Christian influence. Men who have studied in America or Britain are proclaiming loudly the failure of Christianity, and Bolshevist propaganda is actively carried on by others" (Cook, 1922, p. 171).

Religion was refigured as *mixin*, a recent neologism of Japanese origin created to translate "superstition." For example, in 1924, two Chinese Christians, T. C. Li (Li Ganchen) and R. Y. Lo, published *Pochu mixin quanshu* ("Complete documents on smashing and eliminating superstition," though the title is given in English as "Superstitions: Their Origin and Fallacy"). In the preface, they assert the near-uselessness of the Chinese past: "If we removed the superstition from the [24 dynastic] histories, all that would remain would be bits and pieces" (Li and Lo, 1924, p. 1). On the other hand, "Christianity always takes the eradication of superstition as its duty. During this century, superstition was unable to increase its lawless violence because of Christian attacks on it. If we exert our strongest effort to directly attack the old dens of superstition, we could of course conquer the dead life of superstition" (ibid.). Their compilation covers the following topics: geomancy (*fengshui*), divination, physiognomy (*kanxiang*), falling omens [meteors, comets, etc.] (*chuixiang*), becoming a Buddha, becoming an immortal (*chengxian*, a reference to Daoism), supernatural omens (*yaoxiang*), sorcery (*zuodao*, lit., left-handed or "sinister" ways), heterodox theories (*xieshuo*), and polytheism. As we can see, at this point Buddhism was just an instance of a more general field of superstitious beliefs and practices that had to be wiped out.

The classification *mixin* was a powerful tool in the state's drive to modernize, control resources, and suppress opposition. In November 1929, the state issued "Standards for Preserving and Abandoning Gods and Shrines" (Duara, 1995, p. 108). Other laws included: "Procedure for the Abolition of Occupations of Divination, Astronomy, Physiognomy and Palmistry, Sorcery and Geomancy" (1928), and: "Procedures for Banning and Managing Superstitious Objects and Professions" (1930). Religion is acceptable, superstition is not. The list of what constituted superstition was not consistent: "The worship of certain nature gods, the Daoist line of gods who encouraged the use of charms and magical texts, as well as the practices of popular religion were prohibited" (ibid., p. 109). Feuchtwang discusses *mixin* as "a derogatory reference to lists of activities which always, when they are given, include various kinds of fortune-telling, 'fatalism,' spirit-possession, 'witches,' and 'sorcerers,' festivals, the worship of gods and placation of ghosts, and the building of small temples" (Feuchtwang, 1989, p. 45).[13] Eventually, the state utilized the distinction between religion (*zongjiao*) and superstition (*mixin*) to systematically define dangerous religion as superstition, extend control over more mainstream (or controllable) religion, and still proclaim its adherence to the principle of religious freedom.

A vivid example of the iconoclastic mood of the age, and the opposition of religion and science, is an anecdote from the progressive literary journal *Wenxue* (Literature). In 1936, Yingsan (pseud.) contributed a short essay describing activity at the local tutelary shrine as he goes to school to teach "a class in science" (1983, p. 156). Still thinking of this scene, he begins to teach his class of girls.

> These young women had more or less inherited from their mothers and their society some belief in gods, so I attacked their belief using a logical method. I first asked them, "What are the characteristics of wood, stone, and earth? Do they have senses?" They all gave the correct answer from the *Textbook for Science*. Then I again asked, "What is a bodhisattva made of?"
>
> "The craftsman who makes statues uses wood, stone, and earth and forms it in the image of a person," they again replied in unison.
>
> I then concluded, "If so, why believe in that thing that has no senses?" Most of the children who had always believed in me were accepting.
>
> "But sir! The bodhisattva has incense smoke which gives it spiritual potency," a stubborn child's voice objected.
>
> I therefore lectured her on the principle of how incense smoke could only blacken the wood and stone. Finally everyone ceased to hold any other opinion. I took the opportunity to launch an all-out attack, telling them point by point how gods originated, how they developed, how they have been used by rulers, and how they hinder evolution. Finally, I urged that they must completely wipe gods out of their heads and, raising both hands up to my forehead, I gestured as though tearing the gods out. They all gleefully made the same gesture. The class ended with all arms gyrating.
>
> One class of arithmetic, and this day's work was over. (ibid.)

It is impossible to determine the ancestry of this materialistic argument, since we find it in Western sources from the book of Isaiah onwards, as well as in Medieval Chinese sources.

While some missionaries believed the tide had finally turned against idolatry, many others perceived that "idolatry is scarcely less rampant than it was before the Revolution reforms" (Darley, 1914, p. 12). Before long, missionaries complained that the scientism and tumult in China resulted in a backlash, that "the swing of the pendulum has resulted in a great revival of idolatry" (Couche, 1916, p. 93). J. G. Bird notes "the large increase in the number of temples that are being repaired or erected" (1922, p. 170. See also "A Surprise for the gods!," 1919, p. 12–13).

But as iconoclasm became linked to left-wing Western ideologies—missionaries also wrote of increasing attacks on non-Christian religious institutions: "In this district idols were pulled down and burned by students who allied themselves to the political party. Small temples were pulled down, and the larger temples, after being duly repaired and whitewashed, served as headquarters for the Labour and Peasants' Unions. Conservative Chinese were horror-stricken and indignant at such profane acts of vandalism" (Lee, 1928, p. 7).

Christian sites were not immune from the general lawlessness of the period. As K. M. Griggs wrote in 1929,

> last year the bandits came into the church and put their idols on the Communion Table and then killed the pigs belonging to the Christians and offered them up there with incense to the idols. Ducklings were trampled on and all that could not be taken away was smashed up for sheer wantonness. (Griggs, 1929, p. 155)

Two years earlier, in Hengzhou,

> the "Anti-Landlord League" took possession of All Saints' Church. All landlords had to register under this, or risk forfeiting their rents. For one or two weeks we were unable to use the Church, though no damage was done there. Another Union took possession of St. John's School. (Chapman, 1927, p. 118)

But the situation, like so much in this period, was never clear. According to a 1934 report, "There does not seem to have been any discrimination against Christians or the Church. Temples have been treated worse than churches, and decapitated idols are met with on every road" ("In the track of the Communists," 1934, p. 255).

Eventually, some missionaries evinced a greater fear of secularism than of heathenism. At times the iconoclasm of the modernist students was regretted for aesthetic reasons, as when one missionary commented that "with the least possible thought the Chinese, in their zeal for all that is modern, will demolish their ancient buildings; and many treasures of art and history are lightly discarded" (Boreham, 1936, p. 192). Beyond aesthetic or historical concerns, heathen religiosity itself came to be valued: "it is the keenest idol worshippers we want—they are the ones that will be the keenest worshippers of God" (Griggs, 1923, p. 118). By 1929, Bishop Sing Tsae-seng even appears nostalgic about healthy heathenism:

> The Chinese had a great reputation for their faith in idolatry, but now it is not so. During the last few weeks, in many places in Chekiang, the idols in the temples have been destroyed; in Ningpo city, except the Buddhist and Taoist temple schools, many of the idol temples (such as the temple of the city god), of which there are about seventy-three, are being confiscated and their idols destroyed. Of the seventy-three, forty-two have already been destroyed, the remaining thirty-one are to be examined, whether the idols in them have really done the people any good; if the written records speak of a certain amount of good having been done, they may remain for awhile; if not, they too are to be destroyed. When this destruction began, some of the Christians thought it was a good thing; in reality it is not, because at least the people did reverence and fear the idols; now they reverence neither idols nor the true God, and fear nothing and no one, and there is worse chaos than before, and there is no peace anywhere. (Sing, 1929, p. 95)

Missionaries had often appreciated Confucius even as they railed against Confucianism, but with the "Confucian social order" under attack from so many angles, we find more praise for Confucius as a great man and even as heaven-sent:

> He [God] gave them Confucius, just as He gave to the Hebrew people Moses, to instruct them in the ways of righteousness till He Himself should come. Confucius was a statesman rather than a religious leader, and his teaching is a code of ethics. The Chinese owe a great deal to this remarkable man. (Stewart, 1932, p. 28)

Generally, in the 1930s missionary writings, there was much less antagonism toward non-Christian religion, and even praise for Guanyin as "one great symbol of the merciful and good" (ibid., p. 30). In 1938, a report noted the willing cooperation of Buddhists in running a Red Cross camp on temple grounds ("Co-operation in War Service," 1938, p. 256).

During the World War II coverage of China in the *Church Missionary Outlook* was mostly about medical missions in war zones, and about bombing damage to buildings. In December 1947 we see the return of the hostile rhetoric of idolatry in *Outlook*, significantly in an article about evangelizing: "one member of our team had openly criticized the 'mud god,' who was deaf and dumb and man-made" (Hague, 1947, p. 3). By the very early 1950s, virtually all of the foreign missionaries had left China, and Anglican reportage of Chinese Christianity dwindles to occasional items on Hong Kong or Singapore.

Patterns of communist-inspired iconoclasm

China's history from the collapse of the Qing dynasty in 1911 until the Chinese Communist Party (CCP) came to power in 1949 was particularly chaotic, the result of various shifting conflicts and temporary alliances among the armies of the Western nations, contending warlords, the Guomindang (KMT), the Communists, and the Japanese. During this warfare, many thousands of religious sites were ruined, damaged, or occupied by soldiers under every flag. Some were destroyed during fighting, or by retreating armies to prevent advancing enemies from using them (Myers, 1991, p. 12). This kind of military seizure and destruction applied also to grain supplies, livestock, weapons, vehicles, houses, and indeed people, and hence the destruction of temples by armies does not seem to warrant a special category. However, temples and churches were often singled out for ideological reasons. Let us turn to the specifically Communist forms of iconoclasm in China.

In general, Mao Zedong and other Party leaders accepted the Marxist line on religion. Religion was, on the one hand, a powerful tool of the oppressing classes, and therefore deserved to be attacked. On the other hand, religion was also the expression of the misplaced wishes of the oppressed, and would thus vanish naturally soon after the oppression ended. Religion was considered anti-scientific and anti-modern; China's indigenous religions were remnants of a feudal past which had failed China; and Christianity was tainted by its association with Western imperialism. Mao's vision of peasants spontaneously destroying temples undoubtedly had a basis in fact. He reported in 1927:

> In Liling County, prohibiting superstitious practices and smashing idols have become quite the vogue. . . . In the Lungfeng Nunnery in the North Third District, the peasants and primary school teachers chopped up the wooden idols and actually used the wood to cook meat. More than thirty idols in the Tungfu Monastery in the Southern District were burned by the students and peasants together, and only two small images of Lord Pao were snatched up by an old peasant who said, "Don't commit a sin!" In places where the power of the peasants is predominant, only the older peasants and the women still believe in the gods. (Mao, 1995, pp. 305–6)

Identifying older people and women as the last defenders of religion accords with a common gender construction, of modernity as masculine and virile. But it was not only a few old people and women who objected. Bush cites a number of cases, such as a 1957 Renminribao report from Hebei. "It seems that a cadre entered a temple and removed a Buddha image before which many people made their offerings. Such a large crowd gathered in protest before the gates of the county headquarters building the following day that the county government committee agreed that the image should be restored" (Bush, 1970, p. 322; see also p. 393). In a similar case, in Guangdong, "Mob rioting broke out in one place where a Buddha image was removed, and in another place when it was heard that a small temple was to be destroyed" (ibid., p. 393).

The Communist Party in the years before 1949 had a policy of not antagonizing the peasants unnecessarily. Citing cases in Hunan and Hubei where (in Mao's version of the events) "the landlords exploited the opposition of some peasants to smashing idols," Mao advised:

> It is the peasants who made the idols, and when the time comes they will cast the idols aside with their own hands; there is no need for anyone else to do it for them prematurely. . . . It is for the peasants themselves to cast aside the idols, pull down the temples . . .; it is wrong for anyone else to do it for them. (in Sommer, 1995, p. 306)

Mao's caution could, however, be regarded as merely pragmatic. He certainly did not hesitate to force peasants in other ways. Rather than showing the patient confidence of an assured dialectical materialist, his hands-off policy was influenced by incidents of popular protest against Communist iconoclasm.

Over the years, Party policy fluctuated widely, between direct attacks on religion and a patronizing hands-off approach. The periodization of Communist-inspired iconoclasm roughly follows the major stages of their regime. In a first stage, roughly from the founding of the first soviets in the 1920s to 1949, acts of iconoclasm, confiscation of temple property, and attacks on religious professionals were sporadic. The missionary S. E. Law described how Communist troops that came into Changsha, "entered the American Mission Church of St. John's. They saw the cross and tore it down, stamping on it to show their contempt. They tore all the windows out of the building, burnt the pews and then went round hacking and destroying all the woodwork" (Law, 1931, p. 113–14). However, on the whole, the attacks on religious institutions were usually driven by the need to redistribute land to peasants, and to acquire various forms of capital (grain, buildings, metal) found in religious sites.

A second, overlapping stage in the late 1940s and 1950s, was characterized by widespread and more systematic redistribution of capital, especially land, and by the increased persecution of landlords and other groups deemed "counter-revolutionary." In these circumstances, many temples and churches lost their land, buildings, and personnel. A third stage reached its peak in the mid-1960s but continued throughout the Cultural Revolution (1966–76). This period was marked by widespread violence, virtually a civil war in fact, as zealous young people attacked almost all authority figures and turned on each other. Particularly interesting was a mass campaign to eliminate the "Four Olds" (old ideology or ideas, culture, customs, and habits), in which they destroyed schools, temples and churches, signs, books, and art objects—anything that struck them as antique or nostalgic. Though the frenzy of destruction had tapered by 1970 or so, the political terror could not be resolved until the death of the senile Chairman in 1976. I will discuss these two stages in more detail below.

A fourth stage started in the early 1980s. State policy on religion was changed to allow for the ordination of monks and nuns, the construction of temples, and other limited forms of religious freedom. As the political economy shifted toward a free market model, many city and provincial governments supplied funds for temples to be "restored," at least superficially, in order to create tourist attractions and public parks. In a few cases, Japanese Buddhists funded the restoration of temples associated with the founders of their own lineages (for example, the Qinglong si in Xi'an, where Kūkai studied). Such restorations are not without their price, as for example when an old (wooden) temple is torn down so that a new (concrete) replica of the same temple can be built. In part the enterprise of tearing down old wood and building replicas out of concrete results from limited resources—it is cheaper. But the deeper issue is a working concept of historical preservation very different from the principles underlying Western conservation practices. Whereas Western conservationists attempt to preserve every fragment of the original object and to make restoration work evident in the display (hence, the cracks still show), standard modern Chinese practice has been to hide the cracks or recreate the original from scratch. Viewing a Qing house, dating from 1733, with its leaky roof, peeling paint and termites, Italian conservationist Michelle Cordaro noted that the restoration job should be quite simple, but "Above all, we have to keep them [the Chinese] from 'restoring' it" (in Stille, 1998, p. 42).

Early communist iconoclasm

The predominant patterns of iconoclasm in these two most destructive phases (the 1940s and 1950s, and the Cultural Revolution), are quite distinct. An example of the 1950s iconoclasm is the Dongyue miao (Temple of the Eastern Marchmount) in Xi'an. Founded in 1116, this temple was dedicated to Mount Tai. Its name still appears on some city maps, just inside the East Gate of the Ming-era old city wall. When I tried to find it, however, the Changrenli primary school stood where the temple should have been. I wandered in the alleys until someone told me, "it's in the school." It must have been a large temple complex at one time, judging from the size of the three extant buildings, the *dadian*, *erdian*, and *sandian* (great hall, second, and third halls). However, in the 1950s, it was taken over and ceased to function as a religious site. What images remained inside the halls were smashed before or during the early Cultural Revolution. Now the front half of the compound is a school, there are some dwellings in the middle, and an iron welding yard at the back, occupying the third hall at the rear. The *dadian* and *erdian* are surrounded by ten-foot brick walls, with padlocked doors. The wooden roofs, eaves and doors were in fairly good condition, but there was not much inside of these buildings except dust and a pile of old broken chairs.[14]

The Dongyue miao is typical of many that were occupied in the 1950s by peasant associations, organs of state, or by the People's Liberation Army (PLA). Likewise, the Xiyue miao (Temple of the Western Marchmount), at the foot of Huashan near Xi'an, was occupied by the PLA, who built their homes in the compound. Bill Porter visited in the late 1980s and could not get in: "The

Figure 3.3 Dongyue miao, Xi'an. Photograph by Eric Reinders.

entire site is enclosed by a wall and is off-limits to foreigners. It has been used as a military headquarters and barracks for decades, which is probably why it survived the Red Guards" (Porter, 1993, p. 62). Due to its potential for tourism, the PLA were moved out in the late 1990s and the temple was restored.

Here we encounter once again the irony that so frequently attends iconoclasm: military occupation effectively put and end to the *religious* nature of these sites, but the presence of troops often preserved at least some parts of the formerly religious complex. Whereas some temples were erased from both material existence and memory, these occupations at least preserved an object for the focus of memory. Stationing troops in a site was a deliberate tactic to prevent Red Guard and other attacks. In other cases, city and provincial governments in the 1950s stamped their protective talismans onto a few temples, calling them by such labels as "cultural relic" (*wenwu*) or "unit of civilization" (*wenming danwei*). By affixing the marks of state, and by asserting the site's meaning as cultural rather than religious, many sites were protected. The Daxingshan Temple in Xi'an, for example, was declared a "Shanxi province key cultural relic protected unit" (*Shanxisheng zhongdian wenwu baohu danwei*) in 1955. The Xi'an city government gave funds to restore the temple. (For a similar case see Bush, 1970, p. 326.) Even Zhou Enlai showed some personal interest in preserving the site.

In the 1940s and 1950s a good deal (if not all) of the destruction and confiscation was economically motivated—hence, the government, PLA, or local collectives occupied the compounds and/or redistributed the land to farmers or workers. Monasteries were classed as rich landlords. In addition to occupation of land, some pagodas and temple buildings were dismantled to build new buildings and roads (Bush, 1970, pp. 326–8). In some cases the re-use was phrased in pointedly profane terms. Bush quotes a Guangdong Party leader: "Grave stones, coffins, etc., are used by collective farms for building irrigation works, pig sties, latrines, carts, manure buckets, sheds, small water gates" (ibid., p. 399). Gravestones may have become toilets, but they were still useful.

Christianity had been a target of attacks from the beginning of mission work, due largely to the association with Western imperialism. Throughout the development of Communist power, various groups vented their aggression on Christian property. During the 1940s (in Communist-occupied territories) and early 1950s, the worst hit religious group was undoubtedly Christians.[15] James Myers challenges

> the widespread misperception that most of the violence to organized religion in China was done by the Cultural Revolution which began in 1966. In fact, the truth is far different. By the time the Cultural Revolution broke in the

> late summer of 1966, there was very little of the Catholic Church left to destroy. (Myers, 1991, p. xiv)

Some Chinese Christians protected crucifixes and other religious objects, by putting them into wells, or bricking them up in walls. In order to get their hands on these objects, they sometimes had to join in with looting. In 1947, a mob ransacked the Trappist Cistercian abbey of Our Lady of Consolation, 80 miles North-West of Beijing:

> Soon the crowd turned to the church and its sacristy. Sacred vessels were seized along with the vestments of the priests. It was later learned that Christians had been among the looters, and that the sacred vessels and the best of the Mass vestments had been taken by faithful Chinese Catholics to be hidden for safekeeping. (ibid., p. 7)

I have heard numerous similar cases involving Buddha images. There were both pious and impious motives in the emptying of the chapels, and looting could be a tactic to limit the anti-religious destruction.

However, very few churches remained when congregations re-formed in public in the late 1970s, and very few Daoist and Buddhist temples escaped either. Even such protections as the large Daxingshan Temple could muster were not sufficient, and in 1965 approximately two-thirds of the compound was made into the New Wind Public Park (*Xinfeng gongyuan*), which was administered by the Department of Parks and Forests (*yuanlin bumen*). This seizure continued the established pattern of state occupation, and was accompanied by loud rhetoric against "feudal superstition." However, as the story of the Daxingshan Temple takes us further into the 1960s, "the Buddha images and dharma objects [*faqi*] were completely destroyed" (Wang Yarong, 1986, p. 187).

> Buddhism was banned, monasteries were damaged, rooms and buildings were occupied. Some monks returned to the laity, others scattered to other places. Buddha images, dharma objects and scriptures were destroyed. A small amount was taken into the protection of the Department of Cultural Relics (*wenwu guanli bumen*). When the chaos was over, there were only four monks left in the whole temple. (ibid., p. 143)

Patterns of iconoclasm in the cultural revolution

In the iconoclasm of the Cultural Revolution, the issue was not primarily economic but symbolic. That Revolution was, after all, *cultural*, not economic or military, so they wanted to attack symbols. By that time, however, almost no temples were functioning. Many older people confirmed this to me—the villages used to have lots of small temples, but the religious practices ended in the 1950s rather than in the Cultural Revolution. Nonetheless, the buildings usually remained. A history professor named Zhang recalled a church in his home town that was taken over for use as a stable in the 1950s. It did not matter if people could still see that the building used to be a church. There were no sacred objects or religious services inside. "We weren't allowed even to talk about religion," he said. Then in the 1960s the building itself was torn down, because it still had the distinctive architectural form of a church. In the first case, the issue was to break the perceived power of the church as an institution, absorb its capital, and redistribute it to loyal peasants; in the second case the iconoclasm was economically self-defeating but symbolically compelling.

I do not wish to exaggerate the differences in these two modes, but there were certainly significant shifts in emphasis, a movement from preservation of the material substance (if not the sacrality of the object) toward the obliteration of both material and symbolic objects. Both modes had the intention to harm, but in the first case that intention was usually secondary to the economic imperatives. A good example of the contrast between these modes appears in Gu Hua's story "Pagoda Ridge," which describes the fate of "Our Lady's Temple":

> during the Big Leap Forward a steel-smelting army was billeted in this temple. They pulled down the side halls, using the bricks and tiles to make a furnace and boost the production of steel. Although the next year this nationwide campaign was halted, the temple was taken over by the commune for a huge pig farm, and the goddess and her attendants had to give way for thousands of hogs and sows, Chinese and foreign. In those days, of all livestock pigs were the most prized. But during the hard years when the commune members had no meat, this pig farm was disbanded. In 1966, at the start of the "cultural revolution" to sweep away everything old, the tumbledown temple was razed to the ground. All that remains of it now are the bases of a few pillars, and the slogans written on them—"In Agriculture Learn from Dazhai" and "Never Forget Class Struggle"—so eroded by wind, rain and frost as to be almost indecipherable. (Hua, 1985, pp. 10–11)

An episode in Gao Yuan's account shows a number of related themes already discussed, and it is worth considering in some detail (Yuan, 1987, pp. 93–4). The argument that takes place concerns the categorization of the religious images.

> The next morning, we set out for Dafo Temple like an army of Monkey Kings eager to make havoc under heaven. We found that a company of People's Liberation Army men had moved in during the night to protect the temple. As the Red Guards in my class approached the front hall, which housed a smiling, big-bellied Buddha and four Celestial Kings, one of the soldiers came up as if to greet us. The four pockets on his jacket indicated he was an officer. He held a little red book in one hand. "Little revolutionary generals," he said, "this temple is on the national register of cultural relics. It does not belong to the category of the Four Olds. Please leave and find your Four Olds elsewhere."
>
> Moniter Caolan, who also clutched a little red book, answered him, "Comrade Soldier, how can you say this place doesn't belong to the Four Olds? Don't you see those superstitious monsters?" She gestured toward the Celestial Kings, who indeed looked monstrous, especially the one with a green face and protruding teeth.
>
> "These statues are the work of ancient artisans," said the officer. "They are the wealth of the people. We should not destroy them."

The temple is on the "national register of cultural relics," and "does not belong to the category of the Four Olds." The soldier is repeating here a strategy of redefinition that both protected the material objects in temples, and redefined them as "cultural" rather than "religious." Caolan challenges this categorization, pointing out the "superstitious monsters"—that is, statues of *deva* kings customary in the first hall of Chinese temples. The soldier persists in defining the statues as public capital, and as manifestations of the craft of working people.

> "Can't you see that they oppress the ordinary working people?" Little Mihu said, pointing to a small figure writhing in agony under the foot of one of the fiendish kings.

The vanquished demon figures under the feet of the deva kings represent the enemies of the Dharma in Buddhist iconography. Mihu identifies them with the "ordinary working people." The soldier brushes aside the suggestion:

> "These are figures of legend and have nothing to do with reality," the officer said. Students from other classes had drifted over to hear the debate. The officer seized the initiative. "Come with me," he told the growing audience. "I'll tell you a true story."
>
> Intrigued, we followed him across the temple compound to the hall of the Goddess of Mercy. He led us inside. "You all know that the great Goddess

of Mercy has forty-two arms," he said. "But do you know that only the two big arms in front are real bronze? The Japanese aggressors sawed off the other forty arms and melted them down to make bullets to kill the Chinese people. Japan has a long Buddhist tradition, but they didn't care about that when they needed ammunition. They wanted to destroy the whole statue, but they couldn't uproot it. They finally gave up when the temple elders told them that black water would run out and flood the whole city if the goddess was removed."

"Damn those Japanese devils! Death to the Japanese devils!" several Red Guards cursed.

Here, the idea of the image as miraculously preventing a flood of "black water"—or as capable of causing such a flood—passes without Red Guard comment. Likewise, the acts of destruction (melting down, disfiguring) meted out on the image by the Japanese are the cause for cursing the Japanese, not praising them for their iconoclasm. The Japanese are not praiseworthy even for their attacks on images that oppress the masses. In many Cultural Revolution memoirs one finds that the right to perform iconoclastic acts was also denied to certain "Black" category members: for example, a stigmatized "ultrarightist" teacher, Eclectic Zhu, who leaps up on an old gate with a pickaxe and starts chopping, is immediately called down by Red Guards, who themselves bring down the gate. "Only revolutionaries can take part in this!" (ibid., p. 98). The perceived position of the iconoclast determines the meaning of the act. Iconoclasm was not a generic elimination of certain objects but performances that were themselves symbolic.

The Red Guards, of course, followed a Marxist atheism, and the PLA soldier's attachment to the images is also apparently atheistic. The mutilated image becomes a sign of China's dogged survival in the face of Japanese aggression. We do see a glimpse of the image as an object of power when the soldier reports that the Japanese had ceased their attempts to totally destroy the image through fear of a flood. Even the Red Guards seem to attribute agency of a sort to the images: "Can't you see that they oppress the ordinary working people?" Vestigial though they may be, these references hint at the traditional conception of images as organic beings, but also brings to mind the comments of Gell and Freedberg on the power of images, irrespective of the belief system of the viewer.

The officer then describes the process of making such a big statue: the heaping mounds of earth and the collective effort of bronze smelting. Focusing on the skill of the artisans (rather than esoteric knowledge of design or ritual use), he and the image begin to impress the Red Guards. The awed students "looked at the Goddess of Mercy with new admiration" as the officer asked: "Do you still think we should smash these cultural relics?"

> Just then, we heard a series of crashing sounds. The officer's face fell. He rushed outside, and we ran after him. While he had been holding some of us spellbound before the Goddess of Mercy, other students had regrouped by the four Celestial Kings, tied ropes to them, and pulled them down all at once. The painted clay figures had been smashed into pieces.
>
> The officer looked at the wreckage with tears in his eyes. "You—You—What kind of Red Guards are you?" he stammered. He barked a command and a platoon of soldiers ran over. They linked arms to blockade the hall entrance. The demolitionists retreated. Nobody wanted to clash with the People's Liberation Army. We marched off, waving our red flags, singing the Red Guard song, and exulting in our latest victory over the Four Olds.

This incident notwithstanding, a great number of religious objects were saved from destruction by the PLA, whether by posting guards, boarding up temples, or by removing the objects for hidden storage elsewhere. No doubt the four deva kings received a disproportionate share of the violence, being located in the frontmost hall.

Struggle meetings and erasure of identity

Many people in China remember the early and mid-1950s as a time of relative peace. The wars against warlords, the Japanese, and the Nationalists had subsided, and economic conditions improved. As in the post-war years in America, there was a baby boom. This generation, raised on the rhetoric of revolution, but with no obvious external enemy to revolt against, hit puberty in the 1960s and took at face value Mao's provocative and self-serving slogans: Bomb the Party headquarters! It is right to revolt! The death toll for the Cultural Revolution is unclear, and many speak vaguely of "hundreds of thousands," or "millions." During a 40-day period known as "Red August," the death toll has been estimated at 1,700 for Beijing alone (Wang Youqin, 1999). In some cases, the killing was indirect—caused by cumulative neglect, starvation, and overwork—but in many cases the "struggle sessions" ended with murder.

By the time of the Cultural Revolution, there were few monks and nuns in China, but it is worth examining the violent erasure of personal identity, even if it was not explicitly religious. The parades of demonized bodies, and other activities of the Red Guards, involved the disfiguring or erasure of previous identity, and the re-inscription of a new, imposed identity. This "evil" identity was imposed through the manipulation of signs and through physical embodiment. In practice, attacks on the visible form of identity meant the cutting of hair, hiding the face, painting X's onto the body, removal of jewelry and adornment, to wipe the slate clean for a new inscription. We can also see the creation of blank, generic bodies in the widespread homogenization of clothing. During the Cultural Revolution, Zhai remarks, "Dress also changed. For the next ten years the Chinese wore monotonous dark clothing" (1992, p. 91). Just as a censor takes black ink to colorful prose, homogenous clothing "blacked out" the differences of bodies. The tests of orthodoxy were sometimes quite simple: if you can't get a beer bottle up your trouser leg, it's bourgeois (Jiang, 1997, pp. 30–3). One boy says: "tight pants and pointed shoes are what the Western bourgeoisie admire. For us proletarians they are neither good-looking nor comfortable. What's more, they are detrimental to the revolution, so we must oppose them resolutely" (ibid., p. 30). He and his comrades slash a young man's pants open and cut open his shoes.

Among the list of things the Beijing No. 26 Middle School Red Guards objected to are scents and perfumes, blue jeans, slick hairdos, rocket shoes, tight pants, Hong Kong-style suits, "weird women's outfits," Western clothes, children's watches, actors dressed in strange clothes, bracelets, earrings, longevity chains, and gold pens, along with an assortment of other manifestations of individuality and leisure, such as fences around gardens, resplendent weddings, crickets as pets, finger-guessing games, amusement parks, dirty jokes, nicknames, personal names with "feudal bourgeois overtones," and listing even the names of authors and actors in the credits for movies and plays (in Schoenhals, 1996, pp. 212–22).

Even dead bodies were sometimes homogenized through display and deliberately impersonal disposal. In Wen Jieruo's memoir, the "red armbands" surround her and lash at her with belts, while saying they were being lenient in comparison: "Go look at all the piles of corpses at the No. 1 Middle School. Their eyeballs were gouged out!" (Wen, 1995, p. 708). The bodies of loved ones become a mere substance, heaped in a pile. Wen also recalls the trucks used to remove her mother's body, "piled high with a dozen other corpses . . . lying where it fell when it was heaved up onto the back" (ibid., p. 709).[16]

After a denial or erasure of previous identity, the public redefinition of a body as an iconic object was effected through labels painted or attached to the victims. In order to make the labels stick, the Red Guards used glue, or paint, or wooden plaques hung around the victims' necks. In Jiang Jili's account, a school teacher is struggled against: "Du Hai's mother was standing on a stool, her head lowered to her chest. Two torn shoes, the symbol of immorality, were hung around her neck, along with a sign that read, SANG HONG-ZHEN, OPPRESSOR OF THE YOUNG, DESERVES TEN THOUSAND DEATHS" (Jiang, 1997, p. 144). Or, the father of another student: "He was wearing a tall dunce cap covered with red X's, the sign for a criminal. His wrists were tied together behind his back, and his arms were lifted high behind him. His face had been forced down so that we could not see his face. Around his neck was a heavy wooden sign: CAPITALIST EXECUTIONER SHAN YI-DAN. The name had been written in black ink and crossed out in red paint" (ibid., pp. 150–1). Screamed or chanted, the Red Guard's words of abuse—sometimes recited as the reactionary's confession—served to fix the label, the iconic value, more clearly. The body was thus the physical medium that "made the guilty man the herald of his own condemnation," as Foucault put it (1979, p. 43).

In these two instances, the iconoclastic intention was also embodied: "her head lowered to her chest," "his face had been forced down." Throughout all accounts of Cultural Revolution struggle sessions, there is a persistent postural logic. "We ordered her to lower her head and confess her crimes. She said she hadn't committed any crimes and refused to lower her head. So I pushed her head down" (Yuan, 1987, p. 70). In Jiang's account one hapless victim resists his persecution solely by his posture: "At first Old Qian knelt on the washboard with defiant erectness" (1997, p. 110).

In Gao Yuan's account, the Thought Guards bring the author's father to be criticized. They

> ordered him to kneel down. Papa refused, saying that he had never knelt down in front of anybody, not even his parents, and that he did not know how. The Thought Guards kicked him behind the knees and pulled his hair to force him down. He was held in a jet-plane position for two hours. The rebels took turns holding his arms up like wings and planting their feet on his back. Qin Mao and other rebel leaders made speeches denouncing him as Lingzhi county's biggest capitalist-roader. Finally, they put a feudal-style official's black cap from a Beijing opera costume on Papa's head and then took it off to symbolize his removal from office. (1987, p. 183)

They bring him down and render him immobile, silencing him with threats of violence, the better to make him an icon rather than a person. Finding or making a "feudal-style official's black cap" to symbolize a feudal authority figure was necessary in order to be able to remove it. Names were written only to be crossed out. Thus we sometimes find iconoclasts involved in the production of the objects of their venom. In a remarkable 1974 case described by Jun Jing, villagers (mostly of the Kong clan, claiming descent from Confucius) heard that local Communist officials were coming to tear down the Confucius temple, and pre-emptively dismantled it themselves (Jing, 1996, pp. 54–5). The officials

> had lost an opportunity to dismantle the temple in dramatic fashion to highlight their devotion to the campaign. In retaliation, they brought a selected group of middle-school students to the site of the ruined temple and instructed them to build a gravelike mound from rubble. The children were then led in the shouting of political slogans directed against this makeshift symbol of Confucianism, in a public ritual of humiliation (ibid., p. 55).

The preservers of religious tradition (the Kongs) tear down the temple; the would-be iconoclasts are angry at its destruction and reconstruct it (in part) in order to curse it.[17] The labor of the Red Guards looking for targets was sometimes to make a label stick (to decisively define a body), and sometimes to find or make something tangible (destroyable) to stick the label on.

We have seen how iconoclasm was not always focused only on the image; in general, iconoclasts also attacked the keepers of icons—not always by killing monks, but also by attacking the *habitus* of the monk: indeed, his *habitus* and his habit. The bodies of monks and nuns were treated as wasted resources, expensive to maintain, and better utilized in more productive ways. Likewise, thousands of clerics were "returned to the laity" (*huansu*) in the 1940s and 1950s; few remained in the 1960s. The iconicized body, created by clothing, hairstyle, ritual behavior, and so on, was erased without necessarily killing anyone, by changes in clothing, hairstyle, ritual behavior etc.—by the "microphysics of power" (Foucault, 1979, p. 28). The "Four Olds" to be attacked include one term that is explicitly mental (ideas), two terms that may be considered both mental and bodily (culture, customs), and one term that is strongly located in bodily experience: habits.

Conversely, the ritual specialist's recovery from such bodily iconoclasm required the re-institution of the old habitus, the old "air" of the performance at the same "micro-physical" level—even when the texts were lost and the memories fragmented. In the 1980s in rural China, when State persecution of the descendents of Confucius eased off, many of the Kongs were interested in re-creating the proper ritual bodies of their ancestor Confucius and his living descendents (Jing, 1996, ch. 6). Texts had been destroyed, and many of the last performers were very old. Jun Jing describes how the liturgy was reconstructed communally from fragmented texts, oral memory, and habitual memory in what he calls "memory rehearsals" (ibid., p. 103). In order "to achieve such interpretive unity, these temple managers rehearsed their accounts of the temple's past much as if performers preparing for a play," a play that had not been performed for decades (ibid., p. 62). Jun discusses the retrieval of ritual forms through "habitual memory" (ibid., p. 101), "embodied memories" (ibid., p. 112), or "bodily sedimented memories" (ibid.). The mute quality of this kind of body-knowledge makes it both durable (because it is not in a book that can be burned or in minds prone to forgetting) and fragile (because of the loss when an old hand dies). But if such bodily memories had remained un-performed, if the ritual had gone extinct, we might call it bodily amnesia or ritual amnesia. Such amnesia can be random, accidental, just like some breakage of images; but some such amnesia is systematic, intentional, and imposed.

Conclusion

Along with the meeting of religious traditions was a meeting of anti-religious traditions. A whole series of "anti-" or "counter-" traditions overlapped and interacted. By "counter-religion," I mean discourses and practices against a particular religion that themselves come to take on attributes of a "religious tradition." One important model for Chinese anti-Christians was the long native tradition of anti-Buddhist rhetoric, so that Christianity was not only *perceived* as being only a variant form of Buddhism, but *accused* of it. There were Chinese polemicists (mostly Confucians and Buddhists) who strongly identified themselves as anti-Christian, just as a refutation of Buddhism was an important part of the identity of some Confucians. So too, in their attacks on heathen idolatry, the Protestant missionaries made use of a long Western tradition of polemics against "the Romish Church" dating from the Reformation, and an even longer iconoclastic literature traced back to the Ten Commandments. Hostility to another religion has always been a large part of being religious.

The destruction of an icon will never be complete until the destruction of the worship that created it *as an icon*. By the same token, this embodied "religious consciousness" can survive mass iconoclasm. Visiting a gutted temple in Chengdu, long since looted of icons, William Powell tells of old Chinese women showing their granddaughters how and where to bow in a temple; none of the images were there, but the spatial structure of temples remained intact in the old woman's practice. "The old women were showing the young girls what to do, and indicating where the images along the side-walls had once been. They were also bowing to these remembered images, long since taken away" (Powell, 1984, p. 86).

An erasure of identity is often followed by the inscription of new signs. We find iconoclasts involved in the production of icons in a number of ways: first, they create icons to oppose other icons: the face of Mao was undoubtedly a "religious" icon. Second, iconoclasts sometimes physically construct icons in order to destroy them. The creation of the victim-icon may involve putting on an official's hat on (in order to remove it), dunce caps, cut hair, ink on the face, writing on the body, shoes hung around the neck, or imposed physical postures. Third, they define existing objects as icons (a pair of pants, a gate, a human body). This is the most fluid mode, since the definition may be imposed without consent and spread in a viral manner. Hence we see in memoirs a tremendous anxiety over categories: Black or Red, "cultural" or "religious" or "superstitious."

When the rhetoric of iconoclasm moves so easily between icon and body, it obscures the differences between iconoclasm and murder, and therefore promotes violence. "Iconoclasm is defined by its objects, its 'victims,' the focus of its violence" (Wandel, 1995, p. 26). Or: acts of iconoclasm require first the definition of objects as meaningful victims or targets. We also find the definition of bodies as objects, which allows the signification and hence the violence to proceed without mercy.

4

Ways of Not Seeing: Cultural Redefinition and Iconoclasm

A friend returning from a museum exhibition of Mongolian Buddha images asked one of us why the Buddha's eyes are always closed. At first this perplexed us, because we have never seen a Buddha image with closed eyes. A visit to the exhibition solved the puzzle: the curators had displayed Buddha images just like they display any other piece of medium-sized sculpture—in glass cases, about chest-high, so you can stand there and get a really good look at them. From that angle, the Buddha's eyes indeed seem to be closed, because in fact he is looking down, at where he thinks you ought to be—below him, looking up at him on his high altar. In the originally intended spatial configuration, he would be looking right at you. In a sense, the curators closed the Buddha's eyes while making every angle of his body visible.

In a temple, Buddha images are placed on an altar, and addressed with offering and obeisance. It is obvious that when you remove an image from the altar and put it in a museum, the meaning of the object will be quite different. (This is not only true of images: putting a urinal on display in an art museum changes its meaning too.) Museums are full of objects drawn from religious contexts, and the treatment of the religious meanings varies widely, ranging from an art historical mode emphasizing the strictly aesthetic qualities of the object (the formalist approach) to an ethnographic mode that tries to evoke cultural context. Yet, even in the latter mode, when the image is displayed along with the altar on which it originally stood, the effect is only an enlarging of the object to the level of an "installation." Indeed, what is the status of such a temple inside a museum? It is not a temple, it just signifies one; its attempt to reconstruct the original context for the sacred statues is doomed to fail: can we think of a religious context without religious actions and people in it? Those statues are not sacred after all, they just signify their past (and lost) sacrality.

The only way a museum can replicate the full religious context of the image is by becoming the temple, complete with clouds of incense, candles, worshippers, beggars, a different administrative system, etc. Above all, the physical experience of seeing an icon in a temple is fundamentally different from seeing an identical object in a museum. As Yamaguchi Masao notes, "The space called the museum refused from the beginning to admit the smells and sounds of everyday life" (Yamaguchi, 1991, p. 60). Even the photographs in art historical books about sacred sites are usually depopulated, as if the photographer had asked pilgrims to get out of the way while the picture was taken. However, one cannot chide museums for being inadequate temples; after all, temples are usually inadequate museums.

In this chapter we address the iconoclastic effects of cultural redefinition, as when formerly sacred objects are displayed in secular settings such as a museum. In these cases, even in instances of negative redefinition when the object is displayed as an example of "bad" beliefs and practices, the object itself is preserved, often well taken care of; however, the object de facto ceases being "sacred" and becomes something else. This chapter describes some such processes of transformation. Privileged places to examine cultural redefinition are obviously museums and tourist spots, with all the cultural and discursive practices surrounding them. Donald Preziosi has provocatively defined museum practices as "ways of not seeing" (Preziosi, 1989, p. 70); the same is true of tourism, and of other cultural practices which commodify the gaze. The sacred is something at the same time very evident and very elusive—the key issue is context (both in time and space). A tourist visiting a temple in an exotic locale behaves much like a person visiting a museum: sacred objects (or formerly sacred objects) are seen, and understood within a certain cultural framework, but in ways that at times end up questioning the very core of those objects' sacredness. As a paragon, we also discuss ways in which temples display their icons, as well as borderline cases such as recent temple museums. Since objects on display in museums are no longer sacred, and the reasons for their selection are essentially different from whatever reasons temples might have to display them, from the point of view of the status of sacred objects, there is only a thin threshold separating the registers of benevolent destruction from their malevolent counterparts. In the same way, remodeling and restoration (included the restructuring of the display position) are very close to disfiguring and humiliation—as in the previous example of the Buddha image with his eyes apparently closed.

Buddha images in museums

Exhibitions of religious art are confronted with important problems of meaning. On the one hand, statues and paintings are usually displayed completely outside of their religious contexts, exposed to the public gaze in ways that are almost unthinkable in temple settings, where spatial organization and light often conspire to render those objects barely visible. Despite the obvious nonreligious purpose of museums, however, it is not unknown for people in Japan to bow or make offerings of small coins to some images (this happens more rarely in China or America: see Horton, 2007, p. 9). The criteria for these actions are not very clear. The foci of such ritual attention seem to be primarily the most famous and/or the most "beautiful" among such statues.

Since the meanings of Buddha images are never fixed, stable, or singular, pinpointing the exact nature of the change of meaning is not obvious. As a means to analyze the changes in meaning, we explore some spatial and bodily aspects of this relocation of the image. In more direct iconoclastic acts, the relationship between the targeted images and their viewers is radically changed; acts of iconoclasm change the subject position of the iconoclast vis-à-vis the image. The museum's entire culture of display also creates a specific kind of experiencing the self in relation to the icon. Moving an icon into a museum also affects the physical experience of visitors to museums and temples.

Many changes occur irrespective of the curator's good intentions. Between the same Buddhist image in a temple and in a museum, there are differences in placement (vis-à-vis the viewer), enclosure, labeling, marketing, ritual use, relation to other objects, relation to academic discourses, etc—in other words, differences in the cultures of display. Scholars have variously treated this movement from one culture of display to another as a desacralization or even desecration; as a subtle mode of iconoclasm; as a loss of meaning; as colonialist—or as a reenshrinement into a new "sacred" environment, in the so-called Temple of Culture. St. Mungo Museum of Religious Life and Art, in Glasgow, is an example of pluralistic good will toward religious icons in the museum setting. Yet one of the cofounders, Senior Curator Mark O'Neill, notes the homogenizing and desacralizing effects of secular museums:

> the flat, controlled lighting of galleries transforms paintings into easily comparable units of art. No attempt is ever made to re-create the flickering candlelight of the medieval or Renaissance period, much less add relevant music which would help convey something more of the original "presence" of the works. (1996, p. 192)

O'Neill refers to how well or how poorly museums imitate and evoke the religious context of the object, which is not the same as retaining the actual sacred status of the object. Similarly, there is no attempt to recreate the "flickering candlelight" of Buddhist temples; nor the ugly bare neon lights one finds over altars in many Chinese temples today, nor the ceaseless drone of badly-tuned car engines from the street just over the wall, nor the smell of the backed-up public urinal.

Display strategies

We must allow that display is never only a matter of the gaze, though museums are predominantly oriented to the specifically visual aspects of objects: "What the museum registers is visual distinction, not necessarily cultural significance" (Alpers, 1991, p. 30). Explicitly, museums present objects to the more or less curious gaze, and the other senses are involved more rarely (as in the use of sound-making equipment in children's displays, or rare instances where touching is allowed).[1] The bodily experience of museum objects is conditioned by the flow of "traffic" that is preestablished by the floor plan and its spatial limitations, by signs and arrows, and by the content of the display. The content may be structured on the basis of cognitive sequences such as chronology, geography, or theme. The distribution of objects is both marked (as with paintings, statues, or explanatory signs) and unmarked (as with the frames, blank walls, dividers, fire extinguishers, or security guards). Museums thus create a certain *kinetic persona* in relation to history (as they walk a hundred steps from the Stone Age to modernity), or in relation to global geography (as when "India" has a tenth the floor space of "England"). In an important sense, each display section in a museum is a chronotope with specific relations with other chronotopes constituting the entirety of the museum, in turn related to the ways in which each culture envisions its history and geography. As Donald Preziosi wrote, "the museum is also a theater of geomancy—an ideal topography wherein the proper siting of objects and subjects works to insure the preservation of certain essential values, a certain essential spirit" (Preziosi, 1989, p. 70). Museums encode cultural values into the physical experience of seeing its objects. The gaze is not objective or ideologically neutral. Merleau-Ponty describes the gaze thus:

> In the gaze we have at our disposal a natural instrument analogous to the blind man's stick. The gaze gets more or less from things according to the way in which it questions them, ranges over or dwells on them. To learn to see

> colours is to acquire a certain style of seeing, a new use of one's own body; it is to enrich and recast the body image. (Merleau-Ponty, 1962, p. 153)

It follows, then, that changes in the mechanisms of mediation, the "blind man's stick," necessarily accompany changes in both the object displayed and the viewing self. A change in the mode of display from temples to museums does not directly affect the isolated object touched by the blind man's stick, but assembles a new sequence of objects, and fundamentally changes the stick itself. In other words, museums modify subject positions of both the image and the visitor.

Let us examine some of the ways in which museums reconstrue icons, not only in the labeling, but also in the very physical experience of seeing in comparison with temples—in particular, we will focus on isolation, pointing, the organization of space and its relations with visual perception and movement, and the ideological underpinnings of display.

Isolation and pointing

In terms of museum aesthetics, the image at the heart of a Buddhist temple like Longshansi in Taipei is extremely cluttered, obscured by other objects, and placed at too great a distance from the viewing public. To get a good look at it is difficult, and on busy days, almost impossible. In contrast, many exhibitions today are informed by a spatial logic known as the 'aesthetic hang,' according to which the more space you spend on an object, the more significant it is and the more intense the attention it can command" (Duncan, 1995, p. 98). Carol Duncan wrote: "installation design has consistently and increasingly sought to isolate objects for the concentrated gaze of the aesthetic adept" (ibid., p. 17). Museums point at objects by isolating them. The isolation of art objects is especially pronounced in America; the walls of European museums tend to be more full (also because of the sheer number of objects in their collections).

In both cases—the temple and the museum—there are architectonic constructs that point to the object, but the mechanisms for drawing attention to objects are quite different. Temples typically rely on a series of gates and courtyards along a central axis (a kind of frontal pointing). There is a discernible logic of obeisance and height (vertical pointing); a positioning in the center of balanced pairs of other images (lateral pointing); and a series of layers between interior and exterior ("centering"[2]). The ability to negotiate these distinctions is a combination of material construction and knowledge (*habitus*), which is trained into the body rather than innate. Museums share almost none of these modes of pointing. Part of the power of the "aesthetic hang," a single painting on a huge wall (or a very small painting in a huge frame), is the sense of conspicuous consumption—of the kind, "this object must be important if the museum has excluded everything else from the wall." The blank wall certainly has less content than the temple altar, which consumes conspicuously by filling up rather than emptying. By accumulating other images, golden objects, and offerings, the principal icon is presented as the rich and powerful *center* of the entire space. In a museum, it is the space itself that implies capital; in the temple it is the objects that fill the space. Michael Belcher noted that "many museums try to isolate an object in both visual and contextual terms by placing it in a neutral environment" (Belcher, 1991, p. 147).[3] But as he goes on to say, there is in fact no neutral space. In fact, "It seems very hard to escape the effects of singling out an object and putting it on view. Perhaps one should not try" (Brawne, 1982, p. 34). By being in the museum, objects are always marked, valued, and iconic, even if only "vestigially" so (ibid., p. 21).

But even within the museum, objects are never really "isolated." They are put in relationships. O'Neill describes St. Mungo's museum in Glasgow, where "Isis and Horus share a display with the Virgin Mary with the Christ Child and Guan Yin" (O'Neill, 1995, p. 51), and

> Salvador Dali's *Christ of St John of the Cross* is displayed in the same room as an Islamic prayer rug, a Kalabari Screen from Nigeria, and an Egyptian mummy mask. Very few visitors have found this uniquely heterogeneous mixture difficult to understand or to accept what they have in common—a religious aspiration to transcendent beauty. (ibid., p. 196)

All religions share beauty, of course—the curator sees the aesthetic as the highest religious aspiration.

Space, Visual Perception, and Movement

People move and perceive quite differently in these two sites. Preziosi noted how "the museum situates all objects within viewing spaces that evoke and elicit a proper viewing stance and distance" (Preziosi, 1989, p. 69). Michael Belcher (1991, pp. 112–14) provided diagrams of five basic circulation patterns in museums, but none of them fit, for example, Longshansi, because museums tend to put things where there is space—on all the walls more or less equally, whereas the focus of Longshansi is directed frontally to the image; the same applies to most Buddhist temples in China and Japan. Even though there are other foci of attention, the flow is directed toward the principal image, which however remains mostly obscured or shrouded. Temples also do not construct screens or deliberately produce free-standing "divider" space only for display. Although some museums can be relatively "interactive," most visitors stroll and look quite passively. Museums space demands little of visitors' bodies but a basic mobility and a certain endurance. No floor space in the museum need be set aside for laying out offerings or chanting in large groups, though the rooms should be large enough to accommodate guided tourists and busloads of schoolchildren.

How long do people spend in temples or museums? Time-and-motion studies done at museums in 1924 established nine to fifteen seconds as an average time spent looking at each item in an exhibit; in 1981, the museum anthropologist Nancy Lurie said 30 seconds (Stocking, 1985, p. 10). According to the "exit gradient effect," people linger more near the entrance and speed up as they head toward the exit (Belcher, 1991, p. 112). This may be true of temples like Longshansi, but at present we do not have any time-and-motion studies of temple worship.

Images in museums are usually more visible than in temples, but in both cases, the image is not physically accessible. In the case of the museum, there are ropes, barriers, and the showcase, which Duncan says "give emphasis to the social distance between the privileged spaces into which we look and the public space in which we stand" (Duncan, 1995, p. 68). This statement could certainly be made about Guanyin's space at Longshansi: the image can be seen most clearly only in a postcard on sale there. Some Japanese temples have attempted to solve the problem of low visibility of their icons, now major tourist attractions, by relocating them in specially built "treasure houses"—as we will discuss below. Crowd control is not only a modern problem: the seventh-century Chinese monk Xuanzang cited an Avalokitesvara image protected from the crowds:

> Fearing that the visitors might defile the holy image, the temple-keepers had made wooden railings around the statue at a distance of about seven paces from it. Those who came to worship the image did so outside the railings and were not allowed to go near it. The incense and flowers that they offered were flung to the image from a distance (Li, trans., 1995, p. 105).

Just as there are often certain temple images designated for physical contact—usually related to healing practices—so too the more progressive museums can have objects to touch, especially for children or the vision-impaired. Such items are more or less disposable, replaceable, or at least washable replicas of more guarded objects, or samples of raw materials. Still, in neither case are these hands-on objects the "principal" artifact of display, so the sense of physical separation is again pronounced in both cases.

The pedestals of museums function "to bring an object to a proper viewing height" (Brawne, 1982, p. 86) and this is also true of altars, but "proper height" has different meanings according to the context, and according to the expected physical posture of the visitor. Both museums and temples use showcases for protection of the objects, and these "miniaturised and protected rooms" (ibid., p. 90) also serve as aedicules, defined as "a scaled down architectural enclosure which created for that singled out object its own spatial world" (p. 19).

Labeling

Another important mechanism of pointing is labeling. G. Brown Goode wrote in 1891: "An efficient educational museum may be described as a collection of instructive labels, each illustrated by a well-selected specimen" (Belcher, 1991, p. 156). According to Belcher, museums curators feel "the moral obligation of the museum to identify what it displays" (ibid., p. 153). Labels, taxonomy, and the whole inventory system are techniques of control, ownership, pedigree, and authenticity.

A label seems to "provide context," but this is too simple: aside from any errors or lacuna, the "context" of any object is only partly textual (or not textual at all); so the label has its limits, being unable ever to truly provide authentic context. As Yamaguchi notes, "All exhibitions suffer from the condition of being fake" (Yamaguchi, 1991, p. 67). In other ways, the label goes beyond the limits of the original context of the object, because it tells us things rarely known or thought in the original context, it tells a different story. The label "translates" embodied context into text.

In museums, not very much of the labeling is actually read. Very few people read all of it, and most read only some of it (Belcher, 1991, p. 164). Still, the museum display itself is perceived as a text. In the absence of coercion, Western visitors circulate in the direction they read—from left to right (ibid., p. 112). Coming to a given visual unit (a room, or a large display case), most people start their gaze at the upper left, like a page (p. 192). The presence of strikingly eye-catching objects in the room will disrupt this pattern, but "the actual model for displays is that of the book . . . and the experience of the imagined visitor closest to that of the reader—solitary, introverted, and silent" (O'Neill, 1996, p. 191). Donald Preziosi expands this observation to include the discipline of art history: "the art of art history is inextricably grounded in a logocentric paradigm of signification" (Preziosi, 1989, p. 16).

Museum-style labeling is almost entirely absent from most Buddhist temples in Taiwan, China, and Japan, but many have explanatory signs in the more touristic temples. Temples give away or sell small books about themselves; at some temples there are dedicated volunteers who can tell you about the temple; and there are of course many other labels: above the gates, on the pillars, in front of the icons, and on the walls, for example the common pious couplet: "The Buddha's light shines everywhere; If you ask, there will certainly be a response" (*Foguang puzhao, Youqiu biying*). These ways of speaking about the image can be considered as another set of techniques of control, ownership, pedigree, and authenticity.

Temples and museums, each in their own way, present themselves as caretakers of authenticity. Many writers have noted the religious terminology used to describe museums, as temples of culture, where people go on pilgrimage, to find out "who they are" (O'Neill, 1996, p. 191). Museums, "so resembling and reassembling of the world" (Boon, 1991, p. 256) are, like temples, powerful systems for defining objects, not only in their labels but in their architectonic structures, and in the physical experience of the visitor.

Some objects wobble between the kinds of spaces: Taishō University, a Buddhist university in Tokyo, has its own "object of cult" (*gohonzon*), a statue of Amida classified as an "important cultural property." The icon was originally placed in the Akama temple-shrine complex in present-day Shimonoseki in western Japan; built after the battle of Dan no ura in 1185, it is said to reproduce the face of Emperor Antoku, a child who sank in the sea together with the defeated Taira warriors.[4] Removed by the shrine during the forced "separation of the *kami* from the buddhas" (*shinbutsu bunri*) in the early Meiji period (see Chapter 2), it later ended up at Taishō University. Today it has a very peculiar and ambiguous status as a semi-religious, semi-secular object. It is not the object of any specific worship, but is displayed on important occasions of the life of the university, such as the inauguration ceremony of the academic year. Most of the time, however, it is kept in a tabernacle situated in the back stage of the main hall of the university, and is usually not visible.[5] In other words, it functions partially as a "secret buddha" (*hibutsu*; see below). Also in this case, thus, the statue has been preserved, but its functions are very different from those for which it was originally made. Its official status is that of an art object, but it does also play some limited religious functions; moreover, it is almost never on display.

Missionary collections

A little-studied phenomenon, in which formerly sacred objects are put on display in often derogatory terms, is that of missionary collections. In an account published in 1904, the Anglican missionary T. C. Goodchild stopped at a rest-shed and met an 80-year-old Buddhist monk: "Seeing a large shelf full of idols in the shed, I began to talk to him about the uselessness of worshipping them. He seemed to agree with me" (Goodchild, 1904, p. 11). They talked for a while, and the old man attended a service.

> About a month later, when I went to Ha-p'u again, T'ong Sin-sang [the catechist] said he had a present to hand to me, something from our old priest friend, which I should value very much. He then gave me the old man's priest's hat, his rosary, and two certificates of priesthood . . . with one sweep he gave up all the imaginary merit that he had gained in his life service of the idols. . . . Later on in the day I walked over again to the old man's rest-shed and found that the shelf full of idols had disappeared. He had smashed the mud idols to pieces, and his daughter had swept them into a dustpan and thrown the rubbish out at the back of the shed. (ibid.)

As this anecdote shows, the paraphernalia of non-Christian religion were sometimes destroyed and sometimes given as "trophies" to the missionaries (Burroughs, 1900, p. 188).

In one article, Lloyd described objects he had "gleaned" from a village, emphasizing how filthy they all were: a kitchen god of wood, a scroll of a deity called "Mother," cloth bags with incense from a local temple, and a brass image of the "Goddess of Mercy." How did he come by these objects? He was requested to help a man possessed and in a fever. He went and prayed there, and the other Christians there "prayed that the whole household might renounce idolatry and worship the true God." When the sick man got better, he and his family decided to stop their idolatry, "and the articles mentioned above were brought to me as a proof of their sincerity" (Lloyd, 1900, p. 119). Within the whole system of probation and examination that new converts were subject to, destroying or handing over one's family idols became a measure and an initiatory proof of the sincerity of a Chinese person's conversion, almost a kind of negative baptism. There was a whole hermeneutics of suspicion acting between the missionary and the potential converts, and iconoclasm came to function as a trial by fire.

As "trophies," these images were not always destroyed. Bishop Hoare sent some of the images he had been given back to England, to be displayed at the Church Missionary House (Thompson, 1900). At least one convert was complicit in this display, giving the medical missionary Dr. F. M. Cooper a number of images: "I sent one home, as he begged, for the English to see, that more might be willing to come out and teach his people to worship God" (1924, p. 43). In Taiwan, George Mackay created a private museum, which included "idols enough to stock a temple, ancestral tablets and religious curios, musical instruments, priests' garments, and all the stock in trade of Chinese idolatry" (1896, p. 289). Where did these objects come from? He mentions converts giving them to him as part of their enactment of conversion (ibid., p. 179) (see Figure 4.1).

A QUARTERLY TOKEN FOR JUVENILE SUBSCRIBERS; A GIFT FROM THE CHURCH MISSIONARY SOCIETY.

Buddha.

Who is Buddha? From the picture we see that he is an idol, —one of those false gods that are worshipped by heathen nations,—so well described by the Psalmist when he says, "They have mouths, but they speak not; eyes have they, but they see not; they have ears, but they hear not; noses have

No. 21. April, 1861.

Figure 4.1 From *A Quarterly Token for Juvenile Subscribers* (April 1861), p. 1.

As part of fundraising and recruitment events, missionary organizations displayed the objects of heathenism in England, whether in slide shows or illustrated lectures given by missionaries on furlough, or in exhibitions. The largest such exhibition in England, seen by perhaps a quarter of a million Britons, was the Great Missionary Exhibition of 1909 in London's Royal Agricultural Hall, which recreated a "Chinese street" (as well as an Indian Bazaar and Shinto Shrine). The key element on this street was the "idol temple," with its forlorn Chinese god. It is difficult to evaluate the influence of the missionary traffic in Asian icons, which was probably small compared to the more voluminous trade in icons for private collections and museums. The icons displayed by missionary societies were certainly framed as idols and made visible to a portion of the population especially prone to interpreting them as the paraphernalia of idolatry. But whether as evidence of idolatry or of aesthetic culture, the repurposing of icons certainly deprives these objects or their original sacred status and resignifies them in other terms—either religious (but as instances of bad, fake religion) or secular (art).

Ideology and display

The display of representational objects, sacred or profane, is never neutral, devoid of meaning or ideological significance. Objects put on display are chosen on the basis of representational criteria, broadly determined by political and ideological conditions. The job of an object on display is to be seen. In other words, individual and institutions, by displaying their own semiophores (representational objects; on this concept, see Chapter 5), create a space of visibility (and a virtual one of invisibility, made up of potential objects not on display), organize meaning, mobilize bodies—veritable strategies of subjectification based on "showing" objects. As Preziosi has written,

> any museum, in incorporating selections and silences, is an ideological apparatus. In addition, every museum generates ways of not seeing and inhibits the capacity of visitors to imagine alternate histories or social orders, past or future (Preziosi, 1989, p. 70).

Drawing on Foucault's idea of the panopticon—a construction that makes everything visible from its center, thereby empowering the viewer and creating a distinct mode of disciplinary subjection—Donald Preziosi has argued that museums and art history are also panoptic: "by lashing together several nineteenth-century dreams of scientificity, art history has been paradigmatic of a certain modernist, panoptic sensibility: a factory for the production of sense for modern Western societies" (Preziosi, 1989, p. xvi). Foucault's paradigmatic panopticon was a prison designed by Bentham, and Preziosi draws the analogy from a prison to art history as a discipline or field:

> the observation site in Bentham's circular prison confers upon the observer an invisibility and detachment from the objects of surveillance. The position is clearly analogous to the epistemological and synoptic position of the art historical and critical (or in general disciplinary) subject. (ibid., p. 36)

Collections provide the illusion that

> objects of surveillance are understandable through removal from their social and historical contexts. What is occluded is the fact that what are taken as data are, in fact, *capta*. Such a fragmentation and bracketing necessarily gives priority to the formal, morphological, and surface characteristics of the observed objects. (ibid.)

In the case of Buddha images in museums, the experts' celebration of the objects (their beauty, their craftsmanship) is achieved at the expense of "context." When pressed, it would be difficult to pinpoint what "context" refers to, because it may not be clear what the "original" context was. "Context" is never stable anyway. Nor is it clear what things, drawn from the entirety of the other culture, should be selected for purchase or reproduction if museums wish to evoke better the object's "context." How big is "context"? The display of a Buddha image might be better "contextualized" by its placement on an altar, or with a large photograph of a temple behind it, an entire room can be decorated to reproduce an entire room of an Edo samurai's private worship hall. But even with this expenditure on display, the *next* room is full of Mayan pottery, or headless Greek statues, or paintings by Jackson Pollock.

The logic of the spatial arrangement of museums may be multiple: by geographic religions, by individual artist, by theme, by "school," and certainly by time period: the museum "composes a narrative, historical space made up of episodes" (Preziosi, 1989, p. 69). Museums distinguish between objects and place them accordingly, so that objects placed in a labeled, spatial sequence may "contextualize" each other, as earlier or later, or all from one country, or all Buddhist. Nonetheless, through these same spatial technologies of placement and design, and through the fact of the *collection*, museums inevitably homogenize history. Preziosi remarked: "Everything takes place in the museum in some eternal contemporaneity; all diachrony, all difference, all multivocality is enframed in synchronicity" (ibid.). One should not, however, imagine the art museum as a space of absolute contemporaneity, totally lacking diachrony, or multivocality. The lived experience of visiting a museum and looking at objects always involves duration, but the duration of the visitor, not the objects on display that seem eternal in comparison.

Museums (and, to a certain extent, temples as well), thus, function as mechanisms of identity, subjectivity, history, and culture. "Visual environments orchestrate signification, deploy and stage relations of power, and construct and embody ideologies through the establishment of frameworks of legibility" (ibid., p. 169). It is not surprising, then, that the state has always been very interested in producing and displaying semiophores of itself.

At least in certain senses, one can say that museums, and the attendant discipline of art history, are "iconoclastic," not through a failure to achieve their goals but precisely by succeeding, not as a matter of intentional disrespect but on the contrary with great respect for the objects. No matter what labeling or evocation of the temple context, the fundamental nature of the museum transforms all objects into "visibles" and "legibles" (semiophores) and neutralizes the space around them. Though occasionally people bow to objects on display, museums remove from the viewer any ritual obligation other than the very passive obligation to observe. Though museums carefully preserve and to a greater or lesser extent describe the object's treatment as an icon, the icon is inescapably an ethnographic artifact, as removed from the ritual as a book about the ritual.

A further problem of recontextualizing Buddhist icons is museums' willingness to display violated icons, such as the Gandharan icon of the Buddha on display in the Michael C. Carlos Museum of Emory University. It is slightly larger than life-size, except for its lack of a head. Naturally we regret its headlessness, but after all are not museums full of headless bodies, armless women, and men without genitals? Fragmented bodies are normal in museums. Iconoclasm—specifically beheading—has created a new aesthetic, of headless bodies, but also, an aesthetic of bodiless heads. Today, in museum shops and designer's furniture stores, you can buy *just the head* of a Buddha, not removed from a body but made that way. So the bodiless head is a kind of new image created out of the accidental aesthetic of the headless body. With very few exceptions, the bodiless head is never seen in Asia; if it was seen, it was a bad thing (or a good thing if you were the iconoclast)—but people would not display a headless body or a bodiless head, unless the display was designed as a museum.[6]

The display of defaced Buddhist icons raises further questions. Modern Western treatments of religious icons tend to sanitize and secularize icons, treating them as idols, as symbols, or as art. As we have already seen in Chapter 1, recent scholarship has shown how Buddha images were created and maintained as living presences, produced through consecration rituals, obeisance, and various discursive means. Icons were imagined and treated as organic beings, and the effective life of the icon was always intimately related to regular offerings and obeisance. The production of the image as a living presence was performed in the ceremony of "dotting (or opening) the eyes," an act which formally marked the animation of the object. When temples do renovations or have to move their images, they take out the "soul" from the images through a ritual closing of their eyes, often papering over their eyes as well. Images on display at exhibitions are, therefore, quite literally "dead"—empty receptacles devoid of their sacred essence. Images also "die," as when destroyed or damaged beyond repair. In these cases, ritual forms of disposal, metaphorically based on human funerals, are carried out. Are we committing some kind of sacrilege in displaying very conspicuously violated sacred icons or by not tending to their ritual needs? This depends on whether or not the image was ever desacralized, whether or not its eyes were ever closed, or whether a funeral was ever held for the broken image. We do not know the true stories of most icons in museums.

It is our suggestion that museums function as agents of iconoclasm. True, the material object is preserved—indeed, museums pride themselves on their role in preserving material culture—but there is a certain kind of violence against the *meaning* of an object.

Temple museums

Iconoclastic aspects are particularly evident in the case of temple museums. In recent years, many major temples in Japan have built their own museums, normally called "treasure palaces" (Hōmotsukan or a variant thereof), for their most important images.

These buildings are by no means a recent invention. Buddhist temples have accumulated throughout the centuries numerous objects, of a more or less sacred value; these objects can be considered as constituting a temple's "collection"—not all of which would be displayed to the public (see Takei, 1998, pp. 159–208). Of course, a temple is not a museum, and the criteria for selection and inclusion of an object in a temple's collection are not necessarily the same as those followed by museum curators. For instance, objects in the possession of temples include icons moved there from other temples (Jp. *kyakubutsu*, lit. "guest buddhas"), images used for personal worship by past monks or temple patrons, and donations; of course, some can be of stunning artistic value. In general, however, objects are there more for their sacred power than because of their artistic beauty or value. To borrow Walter Benjamin's terminology, while museums emphasize "artistic value," objects in a temple's collection exhibit their "cult value" (Benjamin, 1968, pp. 225–44 *passim*)—even though it is not always easy to distinguish between these two kinds of value. As such, they are purported to transcend their objectuality, their materiality, to partake in the spiritual and unconditioned realm of the buddhas. Perhaps, the original model of temple collections is the mythological Dragon Palace (Jp. Ryūgū), the residence of the Dragon King Sagara located at the bottom of the ocean, in which all precious things of the world, and notably Buddhist texts and relics, are preserved.

However, present-day treasure palaces present several features that seem to constitute a break with traditional temple collections. The style of these buildings ranges from what we could call Orientalistic (a modern reinterpretation of traditional Japanese buildings), as in the Reihōkan (Palace of Spiritual Treasures) of Kōryūji in Kyoto, to postmodern, as in the museum of the Byōdōin in Uji, Kyoto prefecture. The main objects of worship are also displayed in these new buildings, and their traditional place in a temple hall is now empty. Interestingly, these treasure houses also impose different trajectories upon visitors. The space of Reihōkan of Kōryūji is a large room with images on display on four sides and two benches at the center for visitors to look at them at ease—much as in ordinary museums. Icons however are placed on a somewhat elevated position, so as to preserve the traditional vertical relation between the worshiper and the object of cult. Indeed, some particular icons do receive worship and food and flower offerings. The museum of Byōdōin is a postmodern, New Age-like building that combines display of sacred objects with pictures and videos. The building is generally very dark; only the artifacts on display are lighted. (Spotlighting is a standard museum practice, a form of framing device.) The visitor has to follow a one-way path inside dark corridors interrupted from time to time by luminous objects on display. In the process, one learns about the history of the temple and the characteristics of its treasures, while at the same time being exposed to a representation of Buddhism as a source of light in a dark world. Still other temples, as mentioned before, have built dioramas reproducing very bright mandalic spaces—whose continuity with traditional modes of representation is, to say the least, rather dubious. For example, the

Taiwanese monastery Fo Guang Shan created a display area for Buddha images which blends into a kind of underground Buddhist theme park with dioramas of 500 arhats. Similarly, recently rebuilt Shingon temples in Japan offer three-dimensional mandalas; while the iconography is usually based on traditional models (in fact, their being produced in a "traditional" fashion is extensively advertised), the overall effect is decidedly New Age-like. One such temple in Tokyo, the Fukagawa Fudōdō (Naritasan Tōkyō Betsuin), even presents a large picture of the Buddha Dainichi (Mahāvairocana) on the *ceiling* of the main hall, in what amounts to an unprecedented iconographical device. In fact, Dainichi is an all-pervasive, cosmic godhead, but he is not like the Christian God, abiding in Heaven and looking down at His creation. Here we see how a process we may call "re-traditionalization" (with its emphasis on images produced according to traditional, ritualized models) can be offset by new, originally non-Buddhist strategies of display. Do these images reflect new doctrinal, Buddhological developments, or are they rather indices of the intrusion of elements foreign to Buddhism? It is true that ceremonial altars are often placed in front of these icons also in their new location, and that priests are supposed to perform rituals for them. But this new trend of making images more visible to visitors in such peculiar settings, controlled environments more appropriate for their preservation, also changes the traditional mode of interaction between people and their sacred images. Images (even old, traditional ones) are given different meaning and serve to express different attitudes toward the sacred. This relocation of images is done in order to preserve them and enhance their sacredness, but it results in a radical redefinition of their semiotic and cultural status. Also in this case, iconoclasm is a matter of interpretation. A concrete effect of this new strategy of display is the fact that many visitors now see temple halls only from the outside (since there are no icons in them), and pay homage to icons in their new museal locations, where they are clearly visible from close range. As a consequence, worship is now becoming much more of a gaze on surfaces (the temple's external surface, the icon's appearance) than a complex bodily activity.

And yet, selecting sacred objects and displaying them as part of processes of cultural redefinition (either positive or negative) may at times result in the resacralization of those images and even of the entire religious system associated with them. A case of the resacralizing effect of state run iconoclastic campaigns has been studied by Høbjerg among the Loma in Guinea. As Høbjerg writes,

> finding themselves as the privileged target of the state-authorised iconoclastic movement, the Loma were led to believe that they possessed

> something particularly valuable. In other words, they came to reflect on their own religious ideas and practices, because of an external influence and the reflexive stance produced has a strengthening, not a weakening, effect. (2002, p. 72)

In still other cases, destruction became an incentive for reconstruction and proliferation of sacred images, as in premodern Japan (see Chapter 2). Positive cultural redefinition, such as the rediscovery of an old sacred image as an art piece and its display at times also operates as a catalyst for its resacralization and subsequent worship. A significant case in point is the Ichiji Kinrin Butchō Nyorai of Chūsonji (Hiraizumi, Iwate prefecture, Japan). This particular manifestation of Mahāvairocana, dating back to the late Heian period, may have been the private object of worship of Fujiwara no Hidehira (1122?–87), the third generation of rulers in Hiraizumi (Yiengpruksawan, 1991, p. 336). We do not know the functions of this magnificent image after the end of the Heian period. As Mimi Yiengpruksawan suggests, "It is conceivable that after 1187, perhaps when Hidehira was interred in the Konjikidō, Chūsonji monks locked up his icon as well, and its existence became secret as that of the mummies' of the northern Fujiwara leaders" (ibid., p. 346).[7] In any case, in modern times, probably after the Meiji persecutions, the image lost its ritual functions and was placed in the temple museum (Hōmotsukan) together with other statues no longer in active service. Later valorized as an artistic treasure, it was recognized by the state as an important cultural property. In 1979, however, the temple withdrew the image from regular display and resacralized it as a *hibutsu*.[8] In other words, after a long period of sacred limbo, this image was rediscovered and revalorized, first as an art object, and later as an important focus of religious worship. But then, paradoxically, it was made into a "secret Buddha" and withdrawn from public view except in a few occasions. At this point, a look at the phenomenon of secret buddhas is in order as it constitutes a peculiar and ambiguous strategy of display aimed at enhancing the sacred value of icons but often verging on iconoclasm.

Secret Buddhas (*hibutsu*): Ways of not showing

Certain Buddhist objects that are considered especially sacred and valuable are usually not put on display; they are kept "secret" and invisible. Particularly significant among such hidden objects are the Japanese *hibutsu*, literally "secret, hidden buddhas." Considered as the most sacred icons in a temple, *hibutsu* seem to question their own semiotic status: how can they convey meaning if they are hidden, invisible? Making sense through hiding: this paradox may help us shed new light on the way in which iconoclastic practices, broadly understood, are in fact closely connected with the very nature of the sacred and its cultural functions and representations.[9]

A *hibutsu* is defined as a Buddhist statue that is normally kept closed in a feretory-like shrine, hidden from public view and, at times, worship. It is visible only at certain times (with frequencies ranging from once in a generation to once a month) and in specific ritual or ritualized contexts such as a temple festival, a temporary exhibition at a museum, or contingent upon the presentation of an "offering" to the temple. Some images are never on display, even to the Buddhist priests in charge of their rituals.

It should be noted that most *hibutsu* are not always or completely secret. Access to some of them is limited to certain priests. Others, however, can be seen during public displays known as *kaichō* ("opening the curtain"; we should mention that such displays often do not make the *hibutsu* really "visible," since it can only be viewed from far away, at the end of a dark and crowded temple hall), and as photographs in books, tourist guidebooks, and magazines. In addition, descriptions, more or less accurate, have always circulated concerning the shape and features of secret buddhas. *Hibutsu* can also be "seen" according to other modalities of vision such as the techniques of "visualization" and "recollection" described in Buddhist scriptures. Thus, invisibility was often a means to enhance imagination (another form of vision).

It is not clear when exactly the set of practices associated with *hibutsu* began. Documents tell of sacred images subject to regimes of secrecy as early as the ninth or tenth centuries. Especially powerful icons often seem to have been kept inside closed feretories or behind a curtain, which made them invisible to most people most of the time. We can identify a bifurcation in the attitude toward Buddhist images beginning around the thirteenth century. While most became more open, available, and visible (leading eventually to the images on display at today's museum-temples), some became more secret and, in some cases, de facto invisible. By the fifteenth century, the secrecy of the icons on display began to be advertised. Even though previous visibility restrictions were a matter of fact, now hiddenness, secrecy, and sacred value deriving from them were emphasized; it is perhaps in this context that the term *hibutsu* as it is currently used today began to be employed. In the Edo period (1600–1868), spectacular displays (*kaichō*) of secret buddhas attracted thousands of people. Particularly popular were the Amida triad from Zenkōji in Nakano (central Japan) which claimed to be the first Buddhist image to arrive to Japan, the Śākyamuni from Seiryōji in Kyoto, which was said to be the first Buddhist image ever to be made in India by King Udayana, and other images from the Kantō region. It was at this time that *hibutsu* became a cultural phenomenon combining forms of sacredness with secularized ways of securing economic and symbolic capital. In modern times, many *hibutsu* have come to be considered important "art objects" exemplifying the traditional artistic sensibility of the Japanese, and this newly acquired status has undoubtedly contributed to the fame and prosperity of temples enshrining them. Temples thus advertise the display of *hibutsu* with the hope of drawing attention and receiving more visitors.[10] Yet here we see a paradox. While the advertisement of a *hibutsu* as an art treasure participates on the one hand in its desacralization (it is thus a form of iconoclasm), on the other the recognition of an icon as an art object may actually serve to strengthen its sacred value and even contribute to the "prosperity of Buddhism."

Buddha images underline an inaccessible, mysterious, and fundamentally other dimension of the real, while at the same time constituting (being) the symbolic, bodily, and material aspects of the deities. Nothing better than an invisible image, then, to secretly "display" this inaccessible and fundamentally other dimension of reality. Because of their invisible and secret status, *hibutsu* can be considered "degree-zero buddhas," pure Buddha-ness, Buddha potentiality. We could even say that *hibutsu*, with their invisibility, display the very concept of Buddha-hood—omnipresent but out of sight. The *hibutsu* embodies and reproduces with its own invisibility the hidden, concealed Buddha-nature without reducing its presence and efficacy.

Broadly speaking, there are two different but interconnected modes to relate to Buddhist images, one more secular and one more sacred. On the one hand, images are now closer to images of "art objects" than to religious icons; their pictures are often closer to touristic souvenirs. On the other hand, images still possess a cultic function; they are not just touristic curiosities but real presences endowed with religious efficacy. These two modes are not sharply separated, since touristic souvenirs, reproductions of art objects, and religious implements reinforce each other in generating a sense of a living presence, a magical force embodied in the statue. In this sense, religious images are mediators between the sacred and the profane, between mere objectuality and supernatural forces, between systems of beliefs, ritual practices, and individual behaviors, between materials, bodies, and meanings—mediators and at the same time liminal entities situated at the boundaries of all the above categories. Precisely because sacred images are mediators between these sets of opposite terms, they are never "purely" sacred, and thus they are always part of a process of their own secularization—an iconoclastic process. What differentiates and, at the same time, enhances the role of the *hibutsu* as religious images is their invisible presence—an invisibility that intensifies their sacredness and, at the same time, their sheer materiality.

Moving icons

As we have just seen, icons enshrined at temples were not static; modern museums were certainly not the only ones to change the context of icons. Before the Western model of museums was introduced into Asia, and even before the modern trade in Oriental art led to the spoliation of temples and palaces, icons themselves were mobile, and their mobility was not in itself necessarily sacrilegious or iconoclastic. Chinese and Japanese sources on the creation of the first image of the Buddha by King Udayana tell stories of the movement of that image from India to East Asia. The statue itself, we are told, asked pious monks worried about the decay of the Dharma, to take it to China (according to Chinese sources) and, later, to Japan (according to Japanese sources). In order to speed up the long travel, the monk would carry the statue during the day; but during the night, it was the statue that walked carrying the monk on its shoulder. In Japan, the legends connected to King Udayana's statue were developed into the origin myth of the image of Śākyamuni enshrined at Seiryōji in Kyoto and brought to Japan from China by the monk Chōnen (938–1016) in 985.[11] In this case, it is not clear whether Chōnen purchased an original piece enshrined in a temple or a replica. Another interesting case is a Chinese image of Kannon worshiped at Sennyūji in Kyoto. According to monks at the temple, Tankai brought back the statue from China in the thirteenth century. This statue is said to be an image of Lady Yōkihi (Ch. Yang Guifei), the Tang period beauty loved by Emperor Xuanzong; after she was killed in a court conspiracy, the emperor had a statue of her made for her salvation.[12] In this case, it is possible that Tankai was able to buy the statue at a Chinese temple because it was not an object of worship there—just an old item in a temple's repository. By that as it may, this statue definitely has a very interesting story: it started as the portrait of an imperial concubine, it was resignified as an image of Kannon/Guanyin, and is now a secret buddha.

In modern times, however, such traffic in icons has become loaded with nationalistic ideology (Keefe, 2007). When the Japanese amassed a large amount of important pieces of the cultural heritage of China and Korea between the late nineteenth century and the end of World War II, they were attempting to reproduce the Western imperialistic model. Japan needed the East Asian heritage in its museums to display what it envisioned as its cultural roots, while at the same time to represent East Asian cultures as relics from the past waiting for the civilizing, modernizing mission of the Japanese occupiers (this was, of course, the core of Japanese propaganda). We should not forget, however, that in several cases, the Japanese were able to appropriate those artifacts because many people in their countries of origin no longer cared for them. Thus, it was only after the Japanese attributed cultural value to those objects that they became again relevant to their countries of origin as part of their own cultural heritage. A recent case highlights the difficulties involved in such issues today. A gang of robbers from South Korea stole, between 1998 and 2002, a number of sacred images from three temples in Japan, claiming that they were bringing back these treasures to their motherland. However, at least one of these pieces, the Amida triad dating back to the Koryo period (918–1392), stolen from the Kakurinji in Kakogawa near Kobe, appears to have been held by the temple more than a 100 years before Toyotomi Hideyoshi's invasion of Korea at the end of the sixteenth century. Moreover, the temple itself claims to have been founded by a monk from Korea in the sixth century. To make things more complicated, the stolen Amida triad was sold to a South Korean businessman for four million won (currently, about US$ 3,500), and subsequently donated to a small temple in South Korea; the temple, however, claims that the image has been stolen after the donation (detectives claim that it is possible that the temple might be hiding the artifact in order not to return it).[13]

In other words, a sacred image worshipped at a Japanese temple was stolen, with the claim that it is in fact part of the South Korean cultural heritage (at this stage, the object has already been redefined, from a sacred object into a profane one). Subsequently, it was not given to a South Korean museum, but was sold instead for profit, and only later, it was apparently donated to a temple, which however does not display it—that is, does not offer it to public worship. As we can see, excessive attachment to sacred icons may result in several registers of iconoclasm. We should recall here that the theft of sacred images was a common war practice in medieval India and the Indianized states of Southeast Asia; in wars among states, the winner would steal the main icon of the losing monarch's temple or palace and take it to his own residence (Davis, 1997, pp. 51–87).

Apart from more or less licit appropriation and translation, throughout history, buddha images have been taken on tour. In Japan, for example, temples arranged *kaichō,* "temporary unveilings of images and other religious treasures usually concealed from the eyes of the profane" (Kornicki, 1994, p. 174). There were two kinds, namely, *igaichō*, when the image was not moved, and *degaichō*, when the image traveled, as on a tour. Though these displays of icons outside of their homes were arranged by the religious institutions, it makes little sense to imagine these displays as "purely" religious. Kornicki comments:

> For the spectators, *kaichō* had various attractions at different levels of understanding: they offered rewards for devotion, the thrill of penetrating secrecy and privacy, and the opportunity for entertainment. For the temples there were attractions at two levels. First there was the value of *kaichō* as publicity and sectarian propaganda, and second, the chance to boost income from the public by way of donations . . . It is clear that, although *kaichō* were religious occasions, they grew into something that partook not only of religious devotion but also of secular entertainment. (1994, p. 178)

In China, similar processions take place. For example, during the annual festival to Mazu in Beigang, Taiwan, the main Mazu image remains stationary, while hundreds of Mazu and other images from lesser temples in the region make their pilgrimage to the main temple for annual audience with the principal icon. Later, these mobile icons roam the streets in their palanquins, pausing here and there to look upon (and thereby bless) various places of business. To alert the icon of its duty to inspect and bless these locations, the carriers move the palanquins directly above huge piles of firecrackers, which are then lit. The roar leaves all around deafened and filthy, and probably does minor damage to the palanquin, but it wakes up the dozing icon.

Asian Buddhists (and Daoists and Hindus as well) also take their icons out, put them on display in processions related to the assertion of territoriality (Grapard, 1994). While the movement of an icon is not necessarily a demotion or even a significant redefinition, certain rearrangements of icons could also serve those purposes. A change in display could shift the meanings of the icons, and could take place within the altar area itself, as in the following case of the buddha images in the Sōkenji.

Oda Nobunaga (1534–82) is well known as the first of the three feudal lords (followed by Toyotomi Hideyoshi and Tokugawa Ieyasu) who endeavored to create a more or less unified polity controlling most of the Japanese archipelago between the end of the sixteenth century and the beginning of the seventeenth century. Nobunaga was very aware of the power of sacralization and the role of religion in the legitimization of political and secular authority—as well as of religious institutions' own secular authority. Among his main enemies in his hegemonic project were the large and powerful Buddhist monastic center of Mount Hiei, the seat of the Tendai sect, and the Ishiyama Honganji, a temple with strong affiliations and enormous land holdings in central Japan. Nobunaga fought against them to the bitter end, in what turned out to be an enormous blow against Japanese Buddhism in general (see Chapter 2). However, Nobunaga was not against religion per se and used it as a tool to reinforce his own aura. The year he was killed (1582), he built a temple, called Sōkenji (literally, the "All-seeing Temple," a veritable Panopticon temple), where he enshrined himself as the supreme god overlooking everyone in the present and future world. It is possible to see here echoes of Christian theology; Christianity had arrived to Japan a few decades before and was becoming quite popular, also thanks to Nobunaga's support (he used it as a tool to debilitate Buddhist institutions). Nobunaga collected images of buddhas and *kami* from all over Japan and enshrined them in his temple and displayed them to show that all these deities were worshipping his own deified self (Asao, 1970–74, n. 3, pp. 55–7; Yasumaro, 1979, p. 23). The temple was later destroyed, and nothing remains of it today.

Sōkenji was nominally a Buddhist temple and enshrined various sacred images; however, the function of these images was quite different from how they were envisioned by contemporary religion. They were not "objects of worship" (*honzon*) themselves, but they were displayed as "subjects of worship," as it were, all together worshipping the sanctified creator and owner of the temple. In other words, Nobunaga changed the meaning of those images; as a result, buddhas and *kami* were displayed worshiping him, the very destroyer of Buddhist power in Japan. Buddhas and *kami* were thus symbolically subjugated (turned from an object of obeisance to a subject or audience of another icon) in Nobunaga's temple, and practically destroyed on the battlefield (through the defeat of their traditional institutions of worship). What is relevant to our discussion here is the politics of display and the transformation of the sacred signification of the religious objects involved.

Tourism

Similar process of resignification (change in meaning-making) also occur with tourism, another set of discursive practices, typically modern, aimed at looking at the world in particular ways. The growth of the modern travel industry has transformed the world into a kind of museum if not even a theme park, where the tourist follows preassigned paths through the world. The consuming and yet uninvolved gaze moves through an ideal itinerary of treasures or "sights." Tourism and museums operate in different ways, of course: travel is less controlled, indeed less controllable. It is harder to guarantee the total exclusion of the smells and sounds of everyday life (the delays, the beggars, the weather), but still, some control is possible. And in fact, even museums are liable to unwanted disruptions—this is why they hire security guards. While in practice museums have limited control, and tourism even less, a high degree of control is possible at the level of representation. We are conditioned to expect to see certain things on our travels, and from our more or less chaotic experiences of travel, we often craft highly selective pictorial narratives, slide shows that resemble tourist agencies' advertisement (the Great Wall, the Forbidden City, Zen gardens in Kyoto, etc.), except that in our own versions we ourselves are in the picture.

Whereas the museum is a highly contained space, the economic incentives of tourism are able to justify sometime massive transformations of landscapes and towns. Yet the effects of tourism often include a homogenization of experience, as local entrepreneurs provide "accommodations" to the habits and sensibilities of visitors. As Kevin Walsh asserts in his critique of the construction of British "heritage,"

> The heritage then, gives space an identity which is different only in terms of surfaces, spaces rather than places, as it is the heritagization of space which denies the idea of historical processes across time and space. Consequently, there is the promotion, at best, of spaces which people construct as different only through the consumption of heritage pastiche, and at worst, perceive as tourist space, points on a leisure map of the mind. (Walsh, 1992, p. 137)

Good examples of "heritagization" catering to the tourist industry (and to tourists' appetites for stereotypes) are the posters welcoming travelers to the Tokyo International Airport in Narita. Their lavish pictures of Mount Fuji, geishas strolling in (Zen) gardens, and landscapes of Kyoto embellished by pagodas present what many tourists look for in Japan. It is interesting to note, though, that very few Japanese, if any, recognize themselves (their cultural identity) in these pictures, whereas many foreign tourists go to Japan searching for other things: tourists from Taiwan and Hong Kong travel to ski resorts in Hokkaido, and many Westerners are attracted by the—equally stereotypical—image of Tokyo as a *Blade Runner*-like metropolis.

While the ideology of heritage tourism claims to bring the consumer into contact with some true and local meaning, tourist sites are inevitably transformed physically and semiotically into a familiar space of consumption. Sometimes the transformations produce jarring tensions between the tourist's experience and the "original" meaning of the site. The older meanings may still be visible, inscribed on the objects on display, and may represent noble ideals for which we can feel only nostalgia, or an exotic otherness belied by the familiarity of the souvenir shops and hotel chains, and by the swarming presence of all the other tourists (Culler, 1988, pp. 153–67).

The Daciensi (Monastery of Great Mercy) in Xi'an, for example, was founded in the Tang dynasty with imperial sponsorship, and has been associated with the famous Tang monk Xuanzang. The temple, and the city of Xi'an, has capitalized on the fame of Xuanzang. Circa 2001, it cost ten *yuan* to get in, and another fifteen *yuan* to walk up the many stairs inside the pagoda. Dayanta, "Big Goose Pagoda," is at the center of the temple and visible as a landmark for miles around. A large statue of Xuanzang now stands majestically outside the main temple gates, positioned with the pagoda in the background, for the benefit of visitors with cameras. The distinctly religious part of Daciensi (that is, where worship of Buddha images takes place) is only a fragment of the temple, one hall, and a paved space in front of it with incense and candle burners. Several large new buildings in traditional temple style (if not materials) have been constructed in accordance with plans to enlarge the site as a religious center. However, most of the space where the old compound must have been is now just open space and shops, and buildings that always seem to be empty and locked up.

Beside the Dayanta is a tourist trap based explicitly on the popular image of Xuanzang, with a big gaudy gate at the front and a bigger, more cartoonish statue of Xuanzang just inside the compound. You are promised a journey to a fantasy world under the ground, but when you go in, it turns out to be a disused underground air-raid shelter, with long sloping corridors and dark, damp rooms to the sides. The redecoration is barely even superficial. In the rooms are various low-budget amusements, each costing another few cents to get in, like a room with fluorescent black-light images of the Buddha.

By contrast, another Tang temple in Xi'an, Qinglongsi (Azure Dragon Monastery) seems authentic. The temple is famous in Japan because Kūkai (774–835) studied there with his master Huiguo. This sectarian connection explains the distinctly Japanese-looking altars and *stupas* dedicated to Kūkai and Huiguo. The buildings look like Tang period buildings, at least to the untrained eye. However, it is all fake. The original temple ceased to exist after about the eleventh century, and was restored just a few years ago with Japanese money. The timber beams visible in the buildings are actually concrete, painted with brown paint and touched up to look like wood.

The situation with Japanese temples is very different. Historical temples are in general very well preserved; Hōryūji near Nara even claims to be the oldest wooden temple in the world. And yet, many temples have become tourist attractions; they have parking lots for cars and buses and souvenir shops, and their compounds have been transformed into sorts of parks, with lawns and paths. Only buildings reputed to be tourist attractions are open to the public, but other halls, in which the real life of a temple takes place (offices, training precincts, the abbot's residence, etc.) are, rightly, off-limits to visitors. Temples are thus double entities: ceremonial sites and tourist attractions—with very little in common between the two. (This, of course, concerns only major temples that cater to tourists; most temples in Japan are in fact family operations providing funeral services for members of their communities and are of no interest to tourists.) A recent book on Mount Kōya highlights such ambivalent nature common to many important Buddhist centers in Asia (Nicoloff, 2008).

Allowing for differences of scale, tourism also deploys museum strategies: isolating the object or site, structuring local space, preconditioning visual perception, restricting physical movement, and labeling and pointing. Tourism is related to museums in offering displays of things, but they are very different in scale: looking at relatively small objects at home, compared to looking at much larger objects like buildings, temples, palaces, and mountains. Tourism requires more physical participation, and is more of a total experience, usually involving differences in food, physical comfort, possible illness, smell, language confusions, and media access. The tourist has a different sense of being connected to events. Yet it is also possible to resist the experience of "being there" by traveling in groups, not talking much to the natives, listening to familiar music through headphones, and by complaining to like-minded fellow tourists. It is also possible for a Romantic adulation for the foreign culture to obscure a tourist's perceptions.[14] In various ways, tourism can certainly increase the sense of separation from the other culture. Bad touristic experiences can harden certain negative stereotypes, now given a greater effect of veracity by the fact of having gone to the place. Many snapshots record not the place, but the physical presence of the camera's owner in that place. Having "been there, done that," the tourist may check the place off the mental list of possible places to visit in the future. The industry of tourism—tour-guides, guidebook publishing, television shows—often reduces cities or countries to short lists of "sights" to see, and focus on directing tourists to a predictable set of specific sites where transportation and shopping are easily facilitated. Other locations are seen only by locals, scholars, and the occasional tourist straying from the beaten track.

A new trend that is taking place in East Asia is the creation of Buddhist-inspired amusement parks, especially in China. It is still too early to establish whether they will function as desacralizing or resacralizing apparatuses, but their fundamental ambiguity is well indicated by a Japanese example—an image of a bodhisattva enshrined in an amusement park. In the Yomiuri Rando amusement park, situated in the western outskirts of Tokyo, there is a Japanese-style garden named Seichi kōen ("Holy ground park") with an octagonal concrete chapel, called Myōkendō, permanently attended to by a nun.[15] This somehow hyperrealistic setting houses an image of the bodhisattva Myōken. The history of this statue, as recounted by Yamamoto Hiroko, is quite remarkable (Yamamoto, 1989, pp. 35–9). The statue was made in 1301 and housed at the Jōmyōji, the family temple of the Watarai priestly clan controlling the Outer Shrine of Ise. As the tutelary bodhisattva of the Watarai clan, Bodhisattva Myōken was part of a complex ritual and ceremonial context. In particular, it appears that the Watarai women were especially involved in its worship, and on the back of the chapel that housed it there was a burial site for the umbilical cords of the newborn babies. The cult continued for several centuries until it was abruptly terminated in 1868 as a consequence of the forced "separation" of Shinto from Buddhism (*shinbutsu bunri*) promoted by the Meiji government. The modern descendants of the Watarai, the Nakanishi House, kept the statue for a few decades, until they sold it to the owner of the Yomiuri Group, a devotee of Myōken. Previously, in 1941, the statue had been designed as "national treasure" (*kokuhō*), but was subsequently demoted in 1950 to "important cultural property" (*jūyō bunkazai*).

This is not a case of iconoclasm in any apparent sense: the statue has been preserved, studied, and reenshrined at a temple built especially to house it. However, if we consider the dynamics of the changes involved, we cannot deny that the statue functions today, in the Yomiuri Rando amusement park, in a very different way from its original location near Ise at the family temple of the Watarai clan. Even though the statue is virtually the same, all of its social, cultural, and ritual contexts (in short, a star deity, whose cult involved Indian, Chinese, and Japanese features, protecting the women of a clan practicing a peculiar and highly influential combinatory religiosity) have been destroyed and are now long forgotten. We can also identify moments of dramatic change: from a functioning deity in a certain medieval context, to a desacralized entity after the Meiji persecution (it was preserved by the Nakanishi family, not by a temple); it was sold as an art object (national treasure first, important cultural property later) and reenshrined at an amusement park (of all places!).

The politics of signs and the "Magic of the State"

With modernization, many Buddhist icons were desacralized and treated as embodiments of aesthetic ideals, rather than presences of the buddhas. Today, some of the most famous Buddhist temples in China and Japan are in many ways close to theme parks, museum establishments, monuments, and "places of memory," rather than (or in addition to) sites of cult or cultivation. There has been pervasive desacralization, even by religious institutions themselves.

There is also a widespread and continuing effort to turn Buddhist icons into art objects, as indicated by the countless publications in which buddha images (*butsuzō*) are treated as mere "sculptures" (*chōkoku*). Even Buddhist priests tend to present icons now not as real presences but as "symbols," "representations" of the buddhas, thus following the secularizing logic of modernity. As art objects they are also associated with national or ethnic identity as exemplars of "heritage" or "national character." Matsuyama Iwao shows well that what we see today as premodern art objects were not originally conceived or used as they are today. In fact, it was since the 1880s that Japanese scholars and government agencies, under Western influence, began to search for Buddhist icons and reevaluate them for their artistic value (Matsuyama, 1993). Ernest Fenollosa and Okakura Tenshin, key figures in this process, clearly ignored or dismissed the sacred value of the icons they studied, treating them instead as embodiment of aesthetic ideals. Both understood centuries-old temple traditions as superstitions and obstacles against the discovery of "art objects." Fenollosa, for example, wrote about his "discovery" of the Yumedono Kannon:

> On fire with the prospect of such a unique treasure, we urged the priests to open it by every argument at our command. They resisted long, alleging that in punishment for sacrilege an earthquake might well destroy the temple. Finally we prevailed. (Fenollosa 1911, vol. 1, p. 50)

Okakura described the same episode in this way:

> Around 1884, during an art history investigation [I carried out] together with Fenollosa and Kano Tetsuya, I approached the priests [of Hōryūji] asking to see the secret buddha [of the Yumedono hall]. The priests replied that if they opened it [the feretory in which it was hidden], thunder would certainly strike [the temple]. . . . But after promising that we would take care of the thunder, the priests agreed to open the door of the hall; they ran away in fear. (1980, p. 36)

It is clear that the "modernizers" and the priests were following entirely different logics. The Hōryūji priests resisted Fenollosa and Okakura's desecrating endeavor because "even though it would be praised as an art object without peer in the world, at that very moment, the Guze Kannon would lose its sacred significance" (Matsuyama, 1993, pp. 177–8). William LaFleur, referring to Watsuji Tetsurō's book, *Restoring the Idols*, explains that the idols thus restored "did not revert to being gods to be worshipped but were now appreciated as works of art" (1990, p. 239). This seems to be a widely accepted view. However, does it present the full picture of the state of Buddhist icons today? Robert Sharf writes:

> There has been much criticism of the "orientalist" penchant for removing holy images from their religious settings in Asian temples and sacred caves, and reinstalling them in museums. The dislocation, we are told, emasculates the power of the image, reducing it to a mere "object of art" to be appreciated for its aesthetic or exotic qualities alone. While I do not want to condone the pilfering of sacred images, I would note that the critique is somewhat misguided. (1999, p. 97)

Sharf roughly identifies the gist of some of our discussion in this book, but in what way are we then misguided? He explains:

> If the museum acts to curtail or restrain the power of sacred icons, so too does the temple. The mistake lies in viewing the icon as a sublime representation of the divine, rather than as a powerful supernatural presence that must be assuaged lest it unleash its power in unforeseen ways. (ibid.)

This is not a mistake we are making, being fully aware (thanks in part to Sharf's writings) of the less Romanticized views of icons. Sharf notes that the move from temple altar to museum pedestal is a redefinition, but also that temples are structures to control the icons. He highlights a lesser-known aspect of icons, namely, that they are objects of power, and not necessarily of benevolent power. Still, viewing the icon as an object of power rather than a representation of power makes it inevitable that the object will lose power when shown to be powerless—an object of manipulation and commerce and secular discourse. After all, the temple was not struck down by lightning or earthquake after Fenollosa and Okakura exposed the hidden buddha, so that while the icon gains value on the art market and/or as a nation's heritage, it loses its sacred power and agency.

When we look at the historical processes that affected the formation of buddha images as modern art objects in Japan, we can identify several phases. At first, toward the beginning of the Meiji period, Buddhist objects were collected, mainly by Western individuals and institutions, because they were no longer considered sacred in Japan: many of their former worshippers desecrated, destroyed, or sold them. Later, in the 1880s, scholars and government agencies, under Western influence, began to search for Buddhist icons and reevaluate them for their artistic value; scholars and institutions generally ignored or dismissed the sacred value of the icons they studied. Still later, since the first decades of the twentieth century, temples themselves began to request official evaluations of their collections, hoping that some objects receive official recognition as "important cultural property" (*jūyō bunkazai*) or even "national treasure" (*kokuhō*). In this way, temples attempted to become more visible and receive more visitors—in this case, the recognition of an icon as an art object might actually contribute to strengthen its sacred value and even to the "prosperity of Buddhism."

In China in the late Qing and the Republican periods, there was massive pilfering by Western adventurers and scholars who used a rhetoric sometimes of protecting the objects, sometimes of simple imperialism. There were also private collectors, for example, missionaries such as previously mentioned George L. MacKay, who created a private museum in Taiwan. Later, in the 1920s–30s, Western and Japanese art historical and museological models were adopted in China under trying circumstances. Warfare and political chaos prevented any lasting establishment of public display. The emperor's treasures were moved in crates around China and eventually to Taiwan, where some of them are now on display. In Taiwan there are some Buddhist museums, such as the underground museum/fantasy land at Fo Guang Shan, with its huge glittering dioramas and an approximation of Western museology. In China, after the 1980s, many new museums and galleries were established, to meet the domestic and international tourist trade, in accordance with a capitalist consumer ethos (see Figure 4.2). The use of categories such as "important cultural property" was probably borrowed from Japan, but the display of plaques attesting to recognition by the government was hardly a new thing—emperors routinely presented plaques to temples.

Yet it is difficult to interpret all of this simply in terms of disenchantment of the world and the "desacralizing logic of modernity." As Michael Taussig has forcefully written, "What exists now is perhaps best thought of as a new amalgam of enchantment and disenchantment, the sacred existing in muted but powerful forms, especially . . . in its 'negative' form as desecration" (1999, p. 13). In his view, desacration is "the closest many of us are going to get to the sacred in this

Figure 4.2 Plaques on Dayanta, Xi'an. Photograph by Eric Reinders.

modern world. Indeed this negative state can come across as more sacred than 'sacred'" (ibid., p. 1). What are the appeals of desecration? Similar to other forms of transgression such as the carnivalesque, the grotesque, or anti-art, desecration has the potential for entertainment and a certain thrill, a sense of being risqué or fashionably dangerous. Desecration may be performed in a self-righteous way, as a performance to show how liberated one is from institutions claiming to define what is normal or right. The provocateur assumes an audience vulnerable to shock value. Of course, the "desecration" of objects no longer considered sacred or valuable seems only quaint, or passé, or psychotic. Desecration is by definition a religious act.

Bernard Faure submits that "desacration . . . is also perhaps incited by the sacred icon itself," as a sign of the "lingering fascination exerted by it, and as a way of recharging things with alienated 'powers'" (Faure, 1998, p. 777). He continues:

> More fundamentally, icons are alive because they have been consecrated. This also explains why they can, and in some cases must, be desecrated. The iconoclast's attempt to destroy the icons is still an acknowledgement of their power, an affirmation qua negation. (ibid., p. 785)

Along these lines, David Freedberg even suggests that the aesthetic mode of thought might be the active repression of an animistic belief in the power of images, so that even (or especially) the most detached art historical commentary is a response to the magical potency of images (Freedberg, 1985, pp. 429–40). This attitude is by no means typically modern. As we have already seen in Chapter 1, Michalski suggested that a certain "fear of images" (*Bildfurcht*) may lay at the basis of instances of Protestant iconoclasm. According to Taussig, defacement, as it reveals the mystery, "may also animate the thing defaced and the mystery revealed may become more mysterious, indicating the curious magic upon which Enlightenment, in its elimination of magic, depends" (1999, pp. 3–4). If this is true, then, iconoclasm and, more generally, the desire to unveil scientific and cultural mysteries, typical of our age, might be continuations in other forms of ancestral animistic beliefs.

The desacralizing logic of modernity, which produced "art objects" out of sacred icons, could never be completely successful. After all, the boundary between art objects and cult objects is not clear-cut. Thus, rather than desacralizing the icons, some were resacralized differently as embodiments of national spirit, as state-icons, by fully exploiting what Michael Taussig calls the "magic of the state." The modern nation state produces, and is based on, a strong form of fetishism, an irrational belief in something higher, all integrating and all-encompassing, a meaning-giving unit, that is nevertheless deeply related to the most basic facts of materiality (Taussig, 1997). It is the modern nation state, as the dominant source of authorized meaning in our historical period, which controls and sometimes encourages resignification with the goal to appropriate for itself past receptacles and symbols of sacred power. In this way, religious art is transformed into embodiments of the aesthetic "spirit" of a nation. For example, imperial palaces or temples in China have been refigured as testimonies to the greatness of the Han Chinese people, with emphasis on the fruits of the labor of the working masses or on some racial/national genius, and emphasis on China's treasures comparing favorably with the Egyptian pyramids, the Vatican, or other Wonders of the World.

Modern art history may strip buddha images of their religious "magic"; this is why we subject them to detailed and detached descriptions of their measurements, materials, iconography, and doctrinal meanings. However, such "scientific" descriptions in most cases end up by giving the statues another magical status. The "spirit" (*tamashii*) of a buddha, taken away from an image through desacration, is replaced by the "spirit" (*kokoro*)—variously defined as national character, or aesthetic sensibility—of another magic entity: the modern nation state. As a very vivid example of this phenomenon, countless books and magazine articles on Buddhist art in Japan bear titles indicating that their contents have something to do with the "heart/soul of Japan" (*Nihon no kokoro*).[16]

Another interesting case of the interchange between religious icons and state icons, as emphasized by a situation of public display, is a petition of Higashi Honganji temple to the Meiji government dated 1871 stating:

> The object of worship (*honzon*) of our sect is Buddha Amida, but that is, respectfully speaking, just a different name for the heavenly ancestral deity of our imperial nation. When he [Amida] appears out of wisdom, he is called Ame no Minakanushi no Mikoto [the first, primordial god in Shinto mythology]; when he appears out of compassion, he is called Buddha Amida. (quoted in Yasumaro, 1979, p. 7)

This is an instance of a religious institution's attempt to "nobilize" its own sacred signs by transforming them into symbols of the state.

As these examples show, the spirit of the buddha inherent in sacred images was never, in the first place, separate from the spirit of power harnessed to sustain political formations and the state in its various incarnations. Buddhist statues were traditionally considered effective for the protection not just of their sponsors and the temples hosting them, but also and in particular, the state. The "magic of the state" is thus always related in complex and inseparable ways to "religious magic." Buddhist artifacts have always been, in the past as in the present, in different ways, symbols of the state. That is why theft and relocation of artworks and relics can generate international incidents, as in the cases of South Korean sacred objects recently stolen in Japan (see above) and of the mummy of the Chan patriarch Shitou Xiqian (700–90) recently returned to China by the Japanese temple that enshrined it (Robson, 2003). In this respect, the history and semantic drift of the concept of "national treasure" (Jp. *kokuhō*, Ch. *guobao*) is significant. In ancient times, "national treasure" was an honorific appellation of the Chinese emperor, later applied to his regalia symbolizing his legitimacy and authority. It was subsequently appropriated by Buddhism used to refer to monks specially designated by the emperor because of their doctrinal knowledge and/or miraculous powers. In recent times, as we have seen, the term came to designate cultural artifacts of especially relevant, though not exclusively religious, value (see Seidel 1981, 1983).

Part Three

Theories

Rethinking the Relations Between the Sacred and Destruction

5

Orders of Destruction: Iconoclasm, Semioclasm, Hieroclasm

Iconoclasm can be configured as a particularly important case of destruction in general. The study of destruction, that is, envisioning destruction as an object of analysis, conjures a limitless hypothetical catalog of the meanings given to destruction and the ways destruction gives meaning to (or takes it away from) things. What possible order might be brought to the study of iconoclasm?

The relations between materiality and the sacred in East Asian cultures have been extremely complex, and therefore so have the meanings of violence against the sacred. What did it mean to destroy objects, deface them, throw them in the river, or put them in an art gallery? How can we conceive the relation of the most direct and obvious modes of destruction (obliteration by fire, for example) and the more contextual modes such as hiding or relocating? More specifically, how can we treat phenomena of destruction of sacred objects—including those cases in which sacredness itself is questioned or denied?

At first, destruction appears to be random and disorganized, as when people commit heedless acts of violence against objects and bodies, and "vandals" exercise their unbridled cruel fantasies. There seems to be little or no order to destruction, indeed one might imagine that destruction is chaos, and therefore inherently un-orderable. However, acts of destruction can be shown to follow cultural patterns, and certainly the interpretations (both by the perpetrators and the victims) of destructive acts can make sense. Those acts, as well as the damaged residue, can be made meaningful in terms of conventional discourses and attitudes. In this sense, at least, destruction, far from being the negation of cultural meaning, is a form of cultural activity.

As such, there are difficulties in preserving boundaries to the usage of the term. In order to prevent the term from having almost any meaning, we have laid out a "catalog of destruction" structured around the degree of damage done to the material object. This catalog suggests a number of points underlying our approach to destruction.

First, our intention is to include a wide range of material affects, with complete obliteration occurring quite rarely. We want to draw attention to the practice of displaying a partial obliteration or changing the object's context. "Attacks" may even leave the object intact but change its status or mode of display to achieve a demotion, redefinition, or humiliation. In this way, we hope that discussion of the shifts in the cultural meanings of objects (even when they are not obviously damaged) will remain grounded in the empirical circumstances of the objects.

Second, we want to highlight (or assert) the parallels of harmful and preservative destruction. There are rich ironies in the interplay of hostile and benevolent treatments. Iconoclasts sometimes preserve what they want to be rid of, and preservers sometimes destroy what they treasure.

Third, the damage or destruction of sacred objects can be acts of piety, sacrifice, communication, provocation, or renovation. Thus we want to explore destruction as a mode of cultural production.

Finally, the movement in the sequence toward more abstract or contextual forms of "iconoclasm" prompts us to consider museum practices that often transform the objects even as they are cherished, and to consider the notion of "semioclasm," the destruction of meaning.

At this point, we shift our attention beyond our catalog of destruction to theoretical approaches to the status of sacred objects, in the attempt to further clarify iconoclastic practices from the point of view of their impact on the semiotic structure of the objects affected. In the second part of this chapter, we begin with a general classification of objects centered on representational objects (*semiophores*). Next, we continue by showing that all objects, even the most symbolically charged, can go through a phase of "junk" or disposability (and this phase can last for a long time), in which their meaning is, as it were, suspended between significance and irrelevance; it goes without saying that sacred objects become disposable (and potentially irrelevant) during iconoclastic times and as a consequence of the diffusion of anti-religious ideas. Then, we introduce the concepts of semioclasm and hieroclasm. We understand semioclasm as the destruction, to various degrees, of the semiotic structure of objects; hieroclasm is the destruction or denial of sacred meaning attributed to religious objects (thus, a special case of semioclasm). Next, we discuss the various ways in which the semiotic structure of religious objects can be affected by acts of iconoclasm; in it, we take into account acts of violence against the bodies of religious specialists as particular cases (and often, particularly hideous cases) of iconoclasm.

Iconoclastic situations

Sergiusz Michalski's previously mentioned redefinition of Protestant iconoclasm as "rites of destruction" is particularly significant, as it connects destructive acts to symbolic formations and belief systems. Michalski writes:

> the decapitation of sculptures—perhaps the most frequent type of damage—and the breaking of their limbs possess an obvious symbolism. As David Freedberg has emphasized, the intention was not always to destroy the entire statue: "The aim is to render images powerless, to deprive them of those parts which may be considered to embody their effectiveness." (1993, p. 76)

Michalski continues: "These damages can be connected with semi-magical beliefs expressing a peculiar fear of images (*Bilderfurcht*) but obvious analogies appear here also to medieval acts of punishment" (ibid.). The analogy of punishing icons and punishing bodies is quite frequent (though on occasion even objects that bear no resemblance to bodies can be "punished"), and the relation between violence on things and violence on people is one of the themes of this book. "Fear of images" is explained by Michalski as "the quasi-magical fear of figures coming to life, fear of the very principle of showing the sacredness in an image that aspires to be a likeness" (ibid., p. 96). In other words, acts of iconoclasm are often "symbolic counterparts of real corporal punishments"—with the "necessity of preserving the damaged statue as a symbol of the punishment experienced or of the humiliation of its authority" (p. 77).

Michalski attempts to identify the fundamental features of the logic of these rites of destruction. They seem to be based on symbolic inversion—what Bakhtin defines as carnivalistic attitudes and the semioticians of culture as "anti-behavior" (the explicit reversal of everyday practices and behavior).[1] It is thus according to this logic, in which "normal" attitudes toward the sacred are suspended and reversed, that, as Michalski writes, "the first targets of attack were the most worshipped images or those famed for miracles," as though the attackers "wanted to show that even the 'great idols' were not capable of an anti-iconoclastic miracle in their defense" (ibid.). Thus, the status of sacred objects (and their human users and practitioners) is willfully and actively reversed from its original, enhanced condition, and mocked and humiliated in a parody of received forms of worship and devotion—often through the use of excremental elements and references to the "lower" parts (in Bakhtin's sense) of the body. It is in this sense that we can interpret several cases indicated by Michalski such as smearing sculptures with filth or with cow's blood, throwing images of Mary into latrines, wells, rivers, and lakes (ibid.).

Iconoclastic acts and discourses always develop within specific *contexts*: religious, political, economic, social, each with its own rules and modalities that need to be addressed and analyzed. We find a complex, open-ended list of variables: agents, situations, contexts, techniques, and uses of symbolism.

When we try to bring some order to the many iconoclastic phenomena, we should notice the inventiveness of iconoclasts. In one chapter alone, Sergiusz Michalski mentions, among the iconoclastic acts he describes, operations such as covering icons with sheets, turning their faces to the wall, burning them at the stake, decapitating them, breaking their limbs, breaking off their beards, running them through with swords, gouging out their eyes, smearing them with "filth" and cow's blood, throwing them into latrines, wells, rivers, and fountains, putting them into a stove, hanging them up on gallows (even upside-down), putting them in the pillory, putting them on trial, animalizing icons with rams horns or a cow's tail, and ducking them into the river as in a witch-trial (ibid., pp. 75–98). The range of behavior is strikingly similar to how victimized humans are abused, as we have seen in Chapter 3 and we will discuss again below. The borrowing of juridical punishments and modes of humiliation underlines the associations of iconoclasm and violence on human bodies. In different times and places, icons and sacred objects have been variously covered with paint, pasted over with banners, fed to animals, smashed to pieces with hammers, knocked off a pedestal with an umbrella, thrown into a river, burnt, chopped up, melted down, dismantled for re-use as building materials, and thrown into excremental places (irrigation works, pig sties, latrines, manure buckets, etc.). And then there are all the less direct ways: selling them off to collectors, or putting them in a museum.

The *agents* of destruction must be determined and studied in their respective roles and according to their specific modalities of violence. In addition to natural causes, temples and religious objects were damaged or destroyed by deranged individuals, rioting peasants, military forces, political authorities, other temples' militias, and, in some cases, even by resident monks. At times, more than one of these agents were involved in the actions. For example, the military could act in the name of political authorities together with popular mobs and/or in collusion with other religious institutions. Peasants fought against exploitation and excessive taxation by the temple owning their lands. By destroying a temple they applied the usual punishment against rebellious peasants (destruction of their house); by doing so, they were declaring the unholiness of the temple (evident from its lack of compassion and its economic exploitation). Political authorities, while not always directly responsible for destruction, were influential in ordering and provoking some of the most conspicuous operations against Buddhism. The military enacted orders from political authorities aimed at punishing rebellious monks and religious institutions. Temples' militias in Japan were deployed against rival temples in disputes over doctrine, ritual, economic interest, and territorial control. All the above cases of destruction challenged the religious status of the objects that were being destroyed and the people who were being defrocked, harmed, or killed. The implication was that if a temple can be destroyed, then it is not a sacred place—or, conversely, since this is not a real temple then it must be destroyed to protect "true" Buddhism against degeneration. Either way, there was a set of assumptions as what counts as sacred and its representations and offshoots (temples, icons, religious specialists, lay believers). Similarly, the relation of destruction and re-classification is defined by Gamboni as "de-arting," that is, "the collectively and historically produced devaluation and disqualification of works of art" that goes along with efforts to remove public statues (Gamboni, 1997, p. 313). For example, a temple bell could have been melted because (i) the temple owning it was considered heretical, (ii) it threatened a new political order, (iii) of economic shortage required to recover its raw material (copper), or (iv) it was considered a waste of wealth carried out by a privileged social group (monks). Analogously, priests could be defrocked, legally punished, harmed, or even killed as heretics, as enemies (or friends) of the state, as economically unproductive, or as representatives of a corrupt feudal or bourgeois social class. In these cases, the actual results of iconoclastic actions might have been the same, but their larger cultural motivations were different.

As is clear from the terms—"heretics," "demonic," etc.—such interpretations of destruction are from the perspective of the religious institution attacking heresy; meanwhile, the "heretics" of the "false" religion, targeted by destroyers, could in retrospect re-interpret the destruction of their material presence as supernatural intervention aimed at purification (as it happened in Japan after Nobunaga burned the Enryakuji and again after the Meiji persecutions). And this interpretation, in turn, could turn out to prevail. The successful interpretation is obviously the one that receives more support and diffusion—even though we can imagine ambiguous situations in which an underground interpretation is added to a dominant one (as in the case of the Kakure Kirishitan, "Hidden Christians" in Japan, who treated Buddhist images as Christian images). In this case we have a "semiomachy" (a struggle between semiotic systems or interpretations), which may not be ultimately solved in favor of one side or another. The victimized groups may make little distinction as to whether violence is exercised by the state or by other religious institutions, and the agents of destruction may be entirely distinct from those who provide the dominant interpretation of those acts of violence.

Bruno Latour has identified five types of "iconoclastic gestures" (Latour, 2002, p. 20) namely, (i) the attempt to "free the believers . . . of their false attachments to idols of all sorts and shapes" out of the belief that "it is not only necessary but also possible to *entirely* dispose of intermediaries . . . to access truth . . . and sanctity." According to this position, "idols . . . [are] nothing but obstacles in the path to higher virtues"; (ii) a critical, sometimes scandalous attitude toward received idols in the attempt to create "a word *filled* with active images, moving mediators." In this context, "The damage done to icons is . . . always a charitable injunction to *redirect* . . . attention towards other, newer, fresher, more sacred images"; (iii) acts "against the image[s] to which their opponents *cling* most forcefully"; (iv) actions of "innocent vandals" such as "restaurators" and "museum curators" with "absolutely no idea that they were destroying anything"; they act "not so much out of a hatred for images but out of ignorance, a lust for profit and sheer passion and lunacy"; (v) an attitude of "doubt[ing] the idol breaker as much as the icon worshippers," a "devastating irony against all mediators" not because of the desire to "get rid of them, but because . . . of their fragility" (ibid., pp. 21–9). Latour also emphasizes the overlapping nature of all of these five gestures, and the difficulty to actually distinguish among them, which is also one of the main concerns of this book.

Finally, it is important to pay attention to the semiotic labor and effects involved in iconoclastic acts.[2] These affected not just the religious persons or sacred objects they targeted, but also their respective signifiers and signifieds. There is an important distinction to be made between denying the sacrality of a particular monk and denying the sacrality of the clergy as a whole, or between a critique of the present state of the clergy as opposed to an ideal model of priesthood, as we shall see below.

Structure of a catalog of iconoclasm

Considering the multifarious factors involved, how do we propose making sense of destruction? A common-sense way to begin—yet one that is immediately problematic—is to make an initial distinction between unintentional destruction and intentional destruction, with perhaps an intermediate category of carelessness or neglect (such as passive responses to decay, where the harm was not intentional but could have been avoided by more conscientious vigilance). "Intention" here is closely related to "agency," in the sense that destruction is done by someone (Table 5.1).

Among unintentional forms of destruction we might consider natural causes (earthquakes, typhoons, insects, accidental fires, exposure to the elements) and wear and tear (erosion caused by natural molecular disintegration or repeated ritual use, accidental damage). Such unintentional destruction of sacred objects has no "author," and does not call attention to itself, though it may be a background reminder of the material fragility of the objects, or a factor requiring systems of cleaning, maintenance, and further production of images. The intermediate category of carelessness and neglect refers, for example, to a fire that might have been prevented from consuming some sacred object, or to a failure to protect an image from the elements, resulting in its crumbling. Institutions and individuals operate under particular sets of priorities and senses of value and are more or less conscious of the need to protect certain objects, while other objects, no longer highlighted, are left to fend for themselves. As we saw in Chapters 1 and 2, Buddhist institutions were attentive to both natural destruction and carelessness, and devised ways to prevent them or, after the fact, to cope with their results.

But is intention, or agency, the best way to consider iconoclasm? These distinctions seem reasonable, and in some cases useful, but certain questions immediately arise. In certain cases and at least retrospectively, even "natural" wear and tear has been assigned meanings (see Chapter 2). Images that had been broken accidentally were even said to have spoken about their damage (Hansen, 1990, p. 55). Phenomena one might consider unintentional can be attributed to the intention of an unseen agent: earthquakes, for example, were sometimes understood as acts of deities, buddhas, dragons objecting to human landscaping, or Heaven dissatisfied with the immoral behavior of the emperor. (Today, even in America insurance companies call them "acts of God.") The "natural" disrepair of Buddhist temples may be considered a result of bad karma, or a sign

Table 5.1 Intentionality of destruction

Destruction

unintentional	neglect, passive response	intentional
no agency or natural agency	quasi-natural agency	supernatural and human agency

of the degeneracy of the age and the coming End of the Dharma. Sometimes, an iconographically incorrect image could not be cast successfully even after several attempts, and its repeated cracking was seen as the Buddha himself refusing to be represented, rather than the flukes of the materials used or poor craftsmanship. Given a belief in supernatural agency, or a sense of natural forces having agency, is it appropriate (in an interpretive endeavor) to call such cases "unintentional?" This is not to say that all destruction was perceived as deliberate (again, see Chapter 2), only that a single event can be considered intentional or unintentional according to one's conception of nature or agency, and according to one's specific intentions in speaking of the damage. More fundamentally, we have very little access to the intentions of iconoclasts, who only in rare cases articulate their desires—and even then, we have reason to be suspicious of their statements, which are often after the fact or contradictory. As Dario Gamboni cautioned, the vocabularies for explaining intentions are not equally accessible, and some groups can articulate reasons while others are attributed with a lack of reasoning—"a fact that should be borne in mind by anyone attempting to classify attacks against art according to the presence or absence of explicit motives and to the nature of these" (Gamboni, 1997, p. 170). This is one of the reasons why in this book we have rejected a narrow definition of iconoclasm that requires an explicit philosophical discourse. To restrict ourselves to the mind (or in fact, to the pronouncements) of the authors of philosophical arguments is to exclude from consideration the vast majority of acts of, and discourses on, iconoclasm. In fact, third-party "intention" is something that inheres not in the act itself, but in the interpretation people give to that act. No theory of iconoclasm necessarily represents how iconoclastic agents think or act, and it would be an error to privilege an elusive authorial intention.

The categories we propose in this chapter represent not a list of discrete actions or intentions but rather a chart of vocabulary for talking about actions—a catalogue of interpretations, but rooted empirically in, and organized around, the degree of damage to the material object itself. Our attempt to account for destruction in theoretical terms is less audacious or pretentious than it may seem. Gamboni warned against over-schematization: "I personally doubt that reducing the situations that we shall be dealing with to a kind of logical grammar and coining a phrase for each type would be of much help" (ibid., p. 20). Yet he goes on to praise the move from finding a fixed definition for a term to examining the conditions of the term's use: for example, not "What is vandalism?" but "Why do people say an act is vandalism?" It is in this spirit that we intend our ordering of categories.

What makes a moment of destruction meaningful does not lie in the essential nature of the object, nor precisely in the "authorial intention" of the agent, but in the discourse that surrounds the object before and after it breaks. In many cases, there may be no such discourse: destroyed objects may be "mute." The meanings of destruction depend on multiple operations of interpretation. The same act of iconoclasm can be perceived in many ways: the damage to a statue eaten by rats or ants may be attributed to natural processes; or to unconscious neglect for which the responsible parties might yet be held accountable; or to supernatural intervention (rats are the reborn form of an angry monk, as we saw in Chapter 2). Analogously, destruction caused by religious zeal (from spoiling an image because of excessive devotional rubbing all the way up to self-sacrifice) can be interpreted in many ways, including as purely destructive acts caused by ignorance and superstition. Termites beheading a wooden icon (*such as the case discussed on pp. 96–7*) may be considered proof of the cult's impotence and absurdity, or as an occasion to reinvigorate the cult. Whereas interpretation can be widely diverse, malleable, and open-ended, the physical evidence is a relatively stable datum—so that, while all sides may agree that the icon's head has come off, there might be no agreement on what it means.

Quite rarely does iconoclasm end in the instantaneous and total obliteration of an object, but we must consider instead a range of material effects. Harm can be evaluated according to the degree of material destruction, ranging from complete obliteration of the object to cases where the material object is preserved but radically redefined so as to erase the effect of its sacrality. Regarding such violence as malevolent, let us consider iconoclasm on a continuum according to the degree to which the physical integrity of the object is affected. (See Table 5.2 on p. 181.)

Starting with total destruction, some objects are completely and irreversibly *obliterated*—objects made of wood, paper, and silk can be burned to ashes, which are then dispersed without trace. Obliteration of the object's substance is often incomplete: ashes can be gathered, or a metal statue can be *melted down* to make coins or weapons. While the form of the object is lost, the material substance survives and may be meaningful when it is re-formed into a new object. The same thing happens when a structure is dismantled and the materials are used for different purposes. Whereas temple bells were often melted down to make weapons, this process is reversible: Toyotomi Hideyoshi, for example, used swords to make his own great Buddha image.

A particularly complex and potent form of harm to sacred objects is *disfiguring*, because it encompasses several related phenomena, and because the disfigured object functions as a sign of both the power and the powerlessness of the sacred object. Objects can be partially damaged, but in some cases these damaged objects are kept, at least temporarily, in view. The intention may be a display of the damage itself. Bruce Lincoln has pointed to the fact that disfigured objects are displayed as "a revelation of the profanity, temporality, and corruption inherent to someone or something" (1989, p. 125). He calls this attitude a "profanophany." Lincoln arrives at this term after considering the exhumation and display (but not the simple obliteration) of nuns' corpses during the Spanish Civil War. Such displays were greeted by iconoclasts because they refuted the notion of moral/physical incorruptibility normally associated with sanctity. We should add, however, that there seem to exist another kind of profanophany, in which disfigured objects are displayed by their worshipers as signs of their survival; this pertains more to the logic of self-sacrifice or martyrdom, in which damages enhance the sacred value of objects and bodies affected. (In such cases, profanophany reverses itself into a hierophany.) Acts of disfiguring are usually highly ritualistic. Where the object is a humanoid image, disfiguring tends to concentrate on the face (especially the eyes) and the hands.

Closely related to disfiguring is *humiliation*, perhaps as an intermediary step toward disfiguring or obliteration. Icons can be cursed, kicked, defaced, and inscribed with labels accusing them of superstition, as, for example, in the Chinese Cultural Revolution. As a form of abuse, *theft* may suggest both the value attributed to the stolen object as well as its indignity; the object is abruptly subject to explicit manipulation by unauthorized agents—even though their authority is a matter of perspective, and in some cases theft can be claimed as benevolent custody. Nonetheless, due to the illegality of the act, a stolen object may effectively "disappear." Similarly, *hiding* is a strategy to remove from view certain objects, that are stored in some other unknown or controlled location. Hiding is of course reversible, when a captive object is liberated or a contained object is discovered.

A more subtle form of harm to sacred objects is what we could call *negative cultural redefinition* or even *iconoclastic preservation*. In these instances the object is preserved intact, and even highly visible, but redefined by its displacement into a new and secular context in which the agents may aim, in certain cases, to give it a clearly negative connotation. For example, the Church Missionary Society of the Church of England used to display a Chinese icon in the lobby of its headquarters in London (see Chapter 4, p. 143).[3]

The sequence just outlined includes a multitude of material effects, usually thought of as malevolent. While it would seem that any action resulting in the obliteration of a sacred object must be committed with the goal of harming it,

Table 5.2 Forms of iconoclasm affecting the material integrity of the object

Irreversible loss of physical integrity of the object ↑	Obliteration without residue
	Destruction with residue
	Disfiguring: Partial destruction, damage
	Humiliation: Abuse, insult, demotion
	Theft
↓ More reversible	Hiding: Burial, disguise, imprisonment
	Negative cultural redefinition

such is not always the case. Parallel to the effects of malevolent destruction are a range of destructive acts with rather similar physical effects yet intending (or perceived as intending) to preserve the object—to preserve the cultural meaning of the object, to transfer it into an imperishable state, or to preserve the sacrality of the class of objects to which it belongs. Still, the idea of preservative violence against objects is somewhat counter-intuitive.

This *iconoclastic preservation* or "sacrificial" reinforcement of the sacred nature of the object or class of objects is carried out in varying degrees of material destruction of the sacred object itself, in a continuum of positions again ranging from *total obliteration* to *redefining preservation*. In other words, an object can be completely destroyed not only in order to repudiate the power or existence of a deity but also as an act of transformation, for example, so that the object is transmitted to the invisible world of the buddhas. We might call sacrifice the destruction of an object as a religious duty that ultimately sanctifies the object, but probably the use of this much-contested term obscures more than it clarifies. Not all intentional destruction of sacred objects can be considered disruptive: some objects are specifically created in order to be destroyed, such as paper money, paper houses and boats for the dead, effigies, certain wooden ritual implements, and sand mandalas. Images of the *zaojun* or "stove lord" (also known as the "kitchen god") are burned up every year, not as some kind of sacrilege but to transport the stove lord to the more ethereal realms, where he confers with Heaven; when he returns, he inhabits a brand new image in the kitchen. Some "worn-out" objects are destroyed and replaced routinely, such as talismans at New Year in Japan. Often, the burning of such objects is intended as a mode of communication or transformation, not of destruction (e.g. see Reader & Tanabe, 1988, pp. 122–3, 195–7) (see Figures 5.1 and 5.2)

Figure 5.1 A booth for disposal of amulets and other apotropaic objects at a shrine in Kamakura. Photograph by Eric Reinders.

Figure 5.2 Detail. Photograph by Eric Reinders.

Objects can be temporarily and/or partially destroyed (*dismantled* and removed from the visual field), with the ultimate effect of a renewal or re-sacralization, or to avoid a more damaging destruction inflicted by hostile hands. The Grand Shrine at Ise, in Japan, for example, is cyclically dismantled, while a newly built replica is inaugurated, and the "old" components are sent to sub-shrines for use in construction. Jun Jing described the dismantling of a *Kongmiao* (Confucius temple) by its own community in order to pre-empt a more degrading iconoclasm by local Communists (1996, pp. 52–5). "Benevolent disfiguring" is also known as "embellishing vandalism" (Gamboni, 1997, pp. 212–31), and refers to cases such as the movements to "purify" church art of the traces of certain periods, sometimes with the intention of removing later accretions to reveal the earlier look.[4] Beautification usually involves some destruction, even if everyone prefers the result. The efforts to restore (that is, tear down and re-create) dilapidated buildings in China have sometimes frustrated Western conservators such as Michelle Cordaro who, in reference to a Qing mansion, remarked: "Above all, we have to keep them from 'restoring' it" (quoted in Stille, 1998, p. 42). As W. G. Sebald wrote, reconstruction in the sense above is "a second liquidation of the . . . past" (2004, p. 68). What remains is an abstract idea of continuity with the past rooted in the material presence of a certain object in a certain place, but past practices, ways of life, world views (the cultural meaning of that object) are all but lost.

Images and relics may be ritually "*humiliated*" in order to provoke or stimulate their religious efficacy or to protest an insult against the religious institution (Little, 1993, pp. 26–30, 123, 138, 147). If the image is ultimately restored to its previous or more enhanced position, the epistemological framework that justifies its use is reinforced. Matteo Ricci was aware of these Chinese practices, and noted the humiliation of images: "They do worship a few idols, but when these do not grant them what they desire, they beat them hard and make peace again with them later on" (quoted in Gernet, 1985, p. 82). Another Jesuit missionary, Louis Le Comte (1655–1728) wrote: "The idol is bound with ropes and dragged through the streets, covered in mud and all kinds of filth" (ibid., p. 83).[5] Similar practices affecting Buddhist images are still performed in some parts of Japan (Ishizuka, 1993). In the end, the icon is shown to have power, even if extreme measures are needed to trigger that power.

Objects may be hidden (buried, stored away), either to protect them from violence or to enhance their sacrality through secrecy. An example of *hiding to protect* is the ritual to bury scriptures in order to transmit them to later generations and thereby counter the decline of the Dharma; there is also the mythic case of hiding (for future revelation) the sutras in the Palace of the Dragon King. An example of hiding to enhance sacrality is the case of *hibutsu* (secret Buddhas), sacred images that are kept out of sight from most people as a sign of their power (see Chapter 4; Rambelli, 2002).

We can summarize parallels between malevolent and benevolent acts as in Table 5.3.

These parallels between ritual actions and violent acts are quite striking. Certainly, the use of the term "iconoclasm" for phenomena that are commonly understood as humility, self-sacrifice, martyrdom, or even repairs, would at first seem to many artificial and rather counterintuitive. However, we need to remind ourselves of the literal meaning of the term, namely, "destruction of icons, i.e., artifacts believed to embody the sacred"; we should also consider that destruction can have different meanings, and sometimes it can be viewed in positive terms. In this sense, at least, use of the term *iconoclasm* helps us identify an obscure substratum of violence that seems to animate many religious

Table 5.3 Parallel registers: Malevolent and benevolent iconoclasm

	With the intention to harm	**With the intention to preserve**
Irreversible loss of physical integrity of the object ↑	Obliteration (e.g. burning wooden icons)	Sacrifice (e.g. burning paper money or images, self-destruction)
	Destruction with residue (e.g. melting down metal icons, materials used for other purposes)	Dismantling (e.g. periodic repairs; materials used for religious purposes)
	Disfiguring (e.g. gouging f ace, beheading, graffiti)	Remodeling, restoration as benevolent
	Humiliation (e.g. toppling onto ground, verbal abuse, profanity)	Humility (e.g. flogging an icon to activate its power)
	Theft (to harm)	Theft (to enhance)
	Hiding (e.g. confiscating images, burying objects)	Hiding (e.g. burying sutras, production of hibutsu, hiding to preserve objects)
↓ More reversible	Negative cultural redefinition (e.g. ideological polemics)	Positive cultural redefinition (e.g. in museums, tourism)

practices. The movement in the sequence toward more abstract or contextual forms of "iconoclasm" prompts us to consider the notion of "semioclasm," the destruction of meaning. Before we do that in the next chapter, here we would like to extend our catalog of destruction in a way already suggested by Bruce Lincoln's discussion of the display of corpses: the human body is perhaps our most "iconic" object.

Iconoclasm and the body

Not all targets of violence against sacred objects are humanoid images, and so we have refused to limit our treatment of iconoclasm to only icons in the strict sense. If sacrality was extended to buildings and territory, we must include them. However, iconoclasts have frequently turned their attention to the human body, but attacks on the human form remain a special case, in their frequency, symbolic potency, and also in their relation to violence on actual, living bodies. In Chapter 1 we have already seen some of the ways in which Buddha images were "alive" in China and Japan, with particular attention to the ways in which the Buddha, Buddhist icons, and monks' bodies were homologized through the application of the logic of non-dualism and a rhetoric of animation. Just as there are parallels in the production of icons and (certain kinds of) bodies, there are parallels in their destruction. At the same time, there is an intimate relation between persecution and asceticism, such as in religious self-mortification as well as in millennarian persecution and desires for martyrdom. We may ask how iconoclasm is related to the killing, defacing, humiliation, or desacralizing of the living human body, especially of the sacralized and iconic bodies of monks and nuns.

Iconoclasm of the body usually involves not only an attempt to remove a kind of body from view but also an attempt to assert different meanings of certain bodies. In a sense, all victims are (or become) iconic. Indeed, some are powerful receptacles of meaning only after their deaths, and because of their deaths—as is especially evident in the case of martyrs, who come to represent religious identities. Attacks on images and attacks on people often show striking similarities, in both practice and rhetoric. Acceptance of mass killing is facilitated by imposing a certain definition upon targeted victims—their culture, their mentalities, and their bodies. We may call this redefinition objectification—quite literally, making people into "things" or sub-humans.

The next section of this chapter is about the relation of attacks on objects and attacks on people. We propose a catalog of violent ways to dealing with religious people's bodies, which runs parallel to the catalog of objects' iconoclasm we just presented. We hope that by studying the relation between violence on objects and violence on people, we may contribute to changing our perceptions of both.

A Catalog of bodily injury

In considerations of the ways in which we perceive objects as "alive"—however un-mystical, materialist, agnostic, or detached such considerations may be—the phenomenon of violence on objects has attracted special attention. The mute passivity of even the most life-like icons is most obviously demonstrated by attacks and offenses against them. Similarly, violence on another person is one of the most common ways that personal agency is experienced and accentuated. Violence on objects highlights paradoxically both the passive thingness of the object—upon which the actor imposes his will, accentuating the contrast of active living agent and passive insentient thing—and simultaneously testifies to the thing as an active agent, as a presence at least powerful enough to provoke such reactions. Iconoclasm attests to the power of icons even as the iconoclast demonstrates the icon's impotency.

To recall some of the points we made earlier, the immanence of Buddha in his images, and Buddhahood embodied in ritual—cases in which Buddha is made visible, tangible, and accessible—is typical of icons. We have emphasized the need to recognize a strong sense of the animation of Buddhist icons, and thus to avoid some modern conceptions of icons that exclude their animated nature (as idols, symbols, or art). We also emphasized the two-way analogy of icons and bodies: icons as bodies, bodies as icons. The idea that icons are animated is relevant for a discourse on iconoclasm, especially when we consider the relations of iconoclasm and violence against the bodies of monks.

It is immediately apparent that our catalog of destruction can be adapted to consider harm to religious specialists and, in particular, attempts at the destruction of sacred *habitus*. (See Table 5.4.)

We find an analogous, and equally problematic, distinction between unintentional and intentional harm to the body. Sacred bodies (living or dead) may be destroyed unintentionally, as with natural causes (disease, accidents) and wear and tear (aging). Illness and accidental injury, however, have often been seen as intentional, that is, caused by an invading spirit or irate ancestor, or caused by the ill person himself (through the mechanism of karma). As in the case of objects, in an episteme that did not draw sharp lines between the natural

Table 5.4 Parallel registers: Iconoclasm of the body

	Persecution (attacks bodies/*habitus* with the intention to harm, usually done to religious practitioners)	Asceticism (attacks on bodies/*habitus* with the intention to preserve, usually done by religious practitioners to themselves)
Irreversible loss of physical integrity of the body ↑	Obliteration (e.g. killing, no remains left)	Sacrifice (e.g. self-immolation, martyrdom)
	Destruction with residue	Destruction with residue (e.g. relics)
	Forced disrobing ('melting down')	Samgha disrobing its own "impure" members
	Disfiguring (e.g. torture, assault)	Self-mutilation (e.g. burning off fingers, scarification)
	Humiliation (e.g. ritualized abuse, demotion, mockery)	Humility (e.g. fasting, penance, obeisance)
	Hiding (e.g. imprisonment, house arrest)	Hiding (e.g. meditational confinement, eremeticism)
↓ More reversible	Cultural redefinition (mockery, polemics)	Cultural redefinition (e.g. reform movements)

world and the powers of deities, seemingly unintentional destruction can also be interpreted as intentional. Conversely, a lack of decay was noted as a miracle: a year or more after the death of some holy monks, their tombs were opened and the bodies were found to be still fresh—in contrast to the relatively profane bodies of most others, which just rot.

In Buddhist terms, death (or any change in fortune) is the result of karma. Though karma may be considered a natural cause, and even as lacking intentionality, it is not random (though it may seem so); ultimately, what you do determines what is done to you. Those with high spiritual attainment may be able to discern the destructive workings of karma in subtle ways. For example, the modern Chinese monk Xu Yun foresaw his death in a dream: "I just saw in a dream a cow trample on and break the Buddha-seal bridge; I also saw the stream stop flowing" (1988, p. 212). Thus, the ambiguities of intentionality and agency discussed in terms of material objects are equally applicable to the destruction of bodies.

Moving on to more intentional forms of bodily harm, caused by direct human agency, we see that, similarly again to the case of objects, harm can be done in various modes, ranging from complete obliteration of the body to its preservation but with a radical redefinition of its cultural position. This bodily iconoclasm may range from the execution of monks (when their remains are disposed off without trace) to a secular or anti-religious redefinition of the monkhood. The bodily equivalent of melting down an icon is disrobing, when the physicality of the body survives but its monastic form (essence?) is lost. In some cases the monk may voluntarily disrobe out of disbelief;[6] this form of self-iconoclasm is obviously not allowed to sacred images, but was carried out by religious institutions (social bodies) in particular occasions, as discussed in Chapter 2. Monks and nuns may be subjected to various kinds of private or public disfiguring and humiliation. Such non-lethal attacks may be intended to demonstrate a willful violation of the sanctity of the clerical body, or may be a stage in a process leading to killing or disrobing. Whereas monks and nuns generally "advertise" their bodily identity through clothing, hairstyle, and place of residence, in other cases victims were ordinary people (lay believers), so meanings had to be performatively fixed onto their bodies before their destruction. This process of fixing iconic meaning onto bodies—that is, the production of a living (but doomed) icon—blurs the lines between murder and iconoclasm. All of these hostilities may be called *persecutions*.

Of course, martyrdom and the persecution of the righteous can be very powerful motive forces in religious institutions and movements. The *Lotus Sutra* contains a prophecy of future persecution (perhaps reflecting persecution of the composers of the text), and threatens grave consequences for abusing the Buddha or his disciples. Whereas abuse of true Buddhism results in millennia of repeated violence on the body of the abuser (in hell), the victimization of the future preacher transforms him into a more holy state: "The cloak of the Thus Come One is the thought of tender forbearance and the bearing of insult with equanimity" (Hurvitz (trans.), 1976, p. 180). Millennial groups especially tend to treat their own victimization as transfiguring and sanctifying (Wessinger, 2000). A paradoxical effect is that persecution appears as a necessary prerequisite for martyrdom—or, in other words, destructive violence is a vehicle for the manifestation of sanctity.

Parallel to attacks on human bodies with the intention to harm are "attacks" with the intention to preserve. With a similar logic to the sacrifice of objects, a body can be completely destroyed in order to be transmitted to the invisible world of the buddhas, as in the act of self-sacrifice. A few forms of self-sacrifice leave behind no trace at all, such as when monks drown themselves and their bodies are never recovered. Self-immolation or cremation usually leave behind some residue of the body, possibly *śarira* (relics), which can be made into objects of cult (Benn, 2007, p. 168). In the present day, it takes nothing more than a camera and an Internet connection to produce endless icons of burning monks. Though suicide is generally condemned, Buddhist biographers authorized certain cases of suicide as spiritually admirable—in defense of Buddhism, or as an offering to Buddha. Also, we find cases of voluntary disfiguring (self-mutilation, fasting, acts of humiliation, and penance), disrobing enforced by monastic institutions to protect the sanctity of its class of bodies (the Samgha expels an impurity), voluntary demotion, temporary displacement (pilgrimage, migration), and forms of hiding (*biguan* or meditational solitary confinement, retreat, periods of atonement, eremitic isolation).[7] Though in various ways these practices remove from sight a sacred body, the ultimate goal is to preserve the crucial element in the iconic bodies of monks, in other words, the "purity" of the Samgha. In the case of self-immolation, the relics that remain may actually increase the sacrality of the (now dead) body—and of the collective body of the Samgha.

Just as certain anti-Buddhists have denied any real difference between clerics and laity, rejecting the boundary between them, so too, various reform movements, led notably by Shinran in Japan and monks such as Taixu in China, have systematically attempted to erase the boundary between cleric and laity, or certain features of that boundary, in the interests of saving Buddhism from sanctimony or irrelevance.

Such attacks on monastic bodies are performed usually by the monks and nuns themselves, with the ultimate intention of increasing sacrality or at least the prestige of Buddhism. Considering the register as a whole, we see that in terms of their physical effects, these actions are parallel to actions taken to harm Buddhist bodies. All of this self-directed violence for the sake of spiritual transformation is precisely the field of *asceticism*.

A particularly ambiguous case of harm done to the body of the practitioners is that of self-mummification, an extreme ascetic practice that was carried out in several places in China and Japan. The earliest occurrences are recorded in seventh-century China, with the Chan patriarch Huineng; in Japan we find records dating to the late Heian period (twelfth century). As far as we know, this practice was carried out especially in temples in northeastern Japan, such as Yudono in the Dewa mountain range, an important site of Shugendō (mountain asceticism). Many of these self-mummified buddhas (*sokushinbutsu*, lit. "buddhas in their very body") have been thoroughly studied. It appears that the ascetics involved, many of whom were in fact not monks but lay practitioners, carried out for many years a diet based on naturally mummifying substances such as pine needles and tree sap, while practicing meditation and austerities. We do not know how this mummifying knowledge was transmitted and from where. In the final stage of the process, the ascetic was entombed in mountains caves (perhaps characterized by an environment conducive to mummification) still alive while in meditation; the entrance to the tomb was hermetically sealed from the outside. After a few years, the door was opened and the body of the ascetic was found mostly intact and mummified, and looked as if he was still immersed in meditation. The body was then translated to a temple where it was enshrined as an object of worship technically identical with a Buddha image. This practice is connected to the myth of the "suspended death" during mediation (*nyūjō*) of Kūkai, the founder of the Japanese Shingon school, meditating on Mount Kōya while waiting for the advent of Miroku (Maitreya), the future Buddha (see Hori, 1962; Faure, 1991, pp. 162–4; Nihon mīra kenkyū gurūpu (ed.), 1993). Self-mummification involves many registers of our catalog, including the production of relics (in this case, a whole-body relic), fasting, and hiding in meditation. It can also be envisioned as an extreme form of self-mutilation or self-immolation, in which the profane self is literally destroyed to leave only the subtle, pure body of a Buddha image. Self-mummification was also affected by forms of cultural redefinition, including the ascetics' redefinition of soteriology, the rehabilitation of former criminals—who, through this extreme form of asceticism expiated their sins and became objects of cult, the valorization of marginal religious sites as centers of the cult to the self-mummified buddhas, and, more recently, as tourist sites and subjects of scholarly research. Furthermore, the fact that ascetics' bodies were turned into icons to be worshipped signals an interesting overlapping between our two catalogs of destruction, the one dedicated to objects and the one for bodies.

In premodern China and Japan, the blurring of the distinction between images and people (especially monks) was the result of Buddhist doctrinal moves such as the homology of "monk—Buddha image—Buddha." That blurring was the result of ritual performances of invoking the presence of the deity in the icon or "seat"; and the result of the construction of temples that placed the Buddha (icon) or the abbot in the position of the primal ancestor and/or the emperor. The most destructive iconoclasm treated images as merely material, like "bricks or tiles," or ingots of metal. In medieval Chinese persecutions, people were also spoken of as raw material, as *stuff*, a discursive strategy preceding and attending the destruction of monastic orders. Hence, another kind of blurring occurred when bodies were treated as raw materials and the rhetoric of melting down images was applied to monks and nuns.

Attacks on icons do not always totally obliterate the objects, and violence against people-as-icons does not always lead to death, but both cases involve stripping away the external signs of identity. This iconoclasm of the *habitus* (whether of a body or an image) sometimes means opposing the traditional conception of images as organic objects of power, and speaking of them as mere bricks, or making them into mere bricks through defacement. Iconoclasm similarly turns a monk into a mere body (the raw material of economic production), or a corpse. In death too, the bodies of victims were homogenized.

The icon as a living body, the body as an icon—these were homologies active in East Asian Buddhism, active even in moments when violence was applied to icon and body in tandem. In order to take the form of commercially usable objects, the bronze buddhas had to be transformed back into raw substance, with no visual face or identity. *Stuff* is mute, but perhaps not entirely. What might be said about the agency of that substance, the bronze itself? Pushing Alfred's Gell (1998)'s ideas further, perhaps we might say that there is scarcely anything that does not have some form of agency, so that the transformations of Buddha into copper bar and further into coins might be considered a contestation of the Buddha with the material body of the Buddha image, just as Buddhist cultivation was premised on a will imposed on the bodily substance that obeys its own dictates, and can be used to craft other desired bodies. The body seems to be a raw substance that imposes its "agency" upon us.

Killing a body-as-icon may take the form of actual killing, or may involve erasure of the symbolism inscribed on the body. The ritualized body is also a material image, so that iconoclastic movements were almost always anti-clerical, and (in a selective way) anti-ritual.[8] Whereas the eighth- and ninth-century Byzantine iconoclasm was anti-clerical, the iconoclasm of the Cultural Revolution destroyed icons, disrobed monks (if they could find any), and engaged in a more pervasive erasure of personal identity, certainly through killing but also by other means.

To remove or hide the markers that enable a body to be recognizable as a member of a distinct category of sacred bodies is also iconoclasm, to the extent that the body under attack represents a type with greater cultural capital. The destruction of Buddha images at Bamiyan in 2001, which certainly fits the straightforward definition of iconoclasm, called attention to the destruction or erasure of certain other bodies under the Taliban regime: bodies covered, people killed, people killed for not covering their bodies. In the case of the Taliban, the erasure of public, bodily identity parallels the iconoclasm of the Buddha images.

Semiophores: Sacred objects, representation, and display

Cultural historian Krzysztof Pomian has attempted a general classification of objects on the basis of their nature and function (Pomian, 1997, pp. 82–3); he distinguishes them into "bodies" (*corps*), "waste" (*déchets*), "things" (*choses*), and "semiophores" (*sémiophores*). For clarity, we propose to reformulate Pomian's terminology as follows: respectively, "physical entities," "waste," "tools," and "representational objects." Physical entities are the raw objects found by humans in their environment; waste refers to objects that humans abandon, evacuate, or destroy because they have lost their use and functionality. The third category, "tools," includes those objects used to change other objects—to "transform the visible appearance or observable properties, or again to modify the localization of other objects" (ibid., p. 82). Finally, to the category of the representational objects, or semiophores, belong all "visible objects invested with signification" (p. 80) that are destined to "replace, complete, or prolong the exchange of words or to preserve their traces, by rendering visible and stable that which would otherwise remain evanescent and accessible uniquely to hearing" (p. 83). As examples of representational objects, we have monument as *lieux de memoire*, statues as memorials, religious icons, art objects, but also books and any other object considered as a text; even a rock, once displayed in a museum, acquires representational functions and thus becomes a semiophore. Accordingly, semiophores are the result of particular semiotic labor: they can originally be mere physical entities, tools, or even waste, reutilized for signifying purposes, or they can be particular objects specifically produced as semiotic devices. This typology is especially relevant for the subject of this book, because iconoclasm can be understood, in an important sense, as turning semiophores and other formerly important objects (ritual implements as tools) into "waste" or back to "physical entities" ready to be re-used in different ways. Pomian's four categories need not be treated as stable and self-contained, for objects move from one to another. A stone is a physical entity, but it can be sculpted into a Buddha-image, thus becoming a semiophore. In another set of transformations, a bronze Buddha-image can, as a consequence of iconoclastic actions, become waste and be "recycled" as a mere "physical entity" as bronze ready to make cannons ("tools").

The most obvious places to find semiophores are collections of all kinds, especially museums. Museums are quintessentially full of semiophores, drawing attention to their uniqueness and particularity while at the same time—in the very act of display—imposing a new code between signified and signifier and thereby blurring all objects into homogeneity. Preziosi wrote: "the art museum orchestrates difference and contradiction into a single visible field" (1989, p. 69). However, Pomian also suggests that the invisible can take many different shapes, one of which is definitely the realm of the sacred.

According to Pomian, "each semiophore is inserted in an exchange between two or more partners and between the visible and the invisible" (1997, p. 83). As Pomian explains,

> the opposition between the visible and the invisible can take many and diverse forms. The invisible is spatially distant . . . It is also temporally distant . . . In addition, it is beyond all physical space . . . in a space structured totally differently. It is situated in a time of its own, or outside any passing of time, in eternity itself. It can sometimes have a corporeity or materiality other that that of the elements of the visible world, and sometimes be a sort of pure antimateriality. (1994, pp. 172–3)

"Invisible" here means such things as the Buddha, God, Justice, America, ideologies, past events, etc. As such, a semiophore "substitutes for something invisible" by showing it, reminding one of it, and by preserving its traces. Spatial distance, remoteness in time, and untouchability are characteristics of the "invisible" realities that semiophores signify. In the quote above, "it is situated in a time of its own, or outside any passing of time, in eternity itself" refers to the transcendent dimension that many cultures attribute to the Invisible as the sacred—or, in other words, to the reifying aspect that all ideologies, religions, and systems of thought tend to produce.

Religious rituals in particular produce semiophores, by displaying sacrificial objects to the gaze of humans and gods, and by transferring objects from the realm of the visible to that of the invisible (Pomian, 1987, pp. 30–7). In order to be considered as a semiophore, an object should be conceived as "acknowledg[ing] . . . the existence of a potential audience" even "in another temporal or spatial sphere," as for example in the case of "objects intended for gods or for the dead": in that case, "objects remained visible [even] after having been physically destroyed, crushed and burned" (Pomian, 1994, p. 170). Religious objects, and images in particular, function

> as intermediaries for this world and the next, the sacred and the secular, while at the same time constituting, at the very heart of the secular world, symbols of the distant, the hidden, the absent. In other words, they acted as go-betweens between those who gazed upon them and the invisible from whence they came. (ibid., p. 171)

In particular, icons "represent [. . .] normally invisible personages, living on the other side of the boundary separating the sacred from the secular" (ibid.). Representational objects in temples and museums are thus involved "in the exchange process [taking] place between the visible and the invisible worlds" (ibid., p. 172). This is precisely one of the functions of Buddhist icons—that of making the invisible visible, to make its presence felt and corporeal. As we have seen, this also applies, albeit in a paradoxical way, to objects such as "secret buddhas" (*hibutsu*).

Sacred objects are obviously semiophores, albeit of a special kind. They are set apart, displayed (sometimes, only to the invisible gaze of the realm of the buddhas) in a special, sacred place, and become mediators between the visible and different orders of the invisible—the realm of the Buddhas, the psychology of the individuals participating in the ritual, ideas about objects, theories of representations, human behavior, and conceptions about culture and tradition. More radically, sacred objects (and buddha images in particular) are semiophores of the numinous and real presence of the deity in this world: in addition to being real presences, they display presence itself. In this respect, we should remember a fundamental semiotic law, first enunciated by Roland Barthes: an object is also a sign of its function (Barthes, 1973, pp. 41–2). For example, a Buddha image, in addition to being a representation and a "living" embodiment of the Buddha, also signifies what people normally do to it and with it in a given culture and historical period. Jean Baudrillard further argued that objects also have an additional, meta-semiotic value, what he calls "symbolic value." In his definition the symbolic is "the utopia that puts an end to the topologies of the soul and the body, man and nature, the real and the non-real, birth and death" (1993, p. 133). It is precisely this symbolic investment transcending dualisms and limitations that contributes to give the Buddha images their "aura" as living entities. The symbolic value given to objects turns objects into something else, namely, into mediators between dualistic realms; they partake of opposite cultural categories and seem to acquire a life of their own (what Marx called "commodities fetishism"). Not all objects—not all semiophores—are usually given what Baudrillard calls "symbolic value"; but when they are, representational objects acquire a particular cultural power as "symbols." When semiophores are displayed and/or used to symbolize ideologies, philosophical positions, ways of life, they become the primary targets of iconoclastic activities.[9]

Re-signification and disposability

The possibility for objects, including sacred objects, to change their status and function, sometimes in a very radical way, alerts us to the need to address the processes of transformation, in particular the one going from "semiophores" to "waste" and vice versa. Michael Thompson's study of semiotic transformations of cultural objects can give us some important hints on how to understand this important aspect of iconoclasm. Thompson distinguishes between two general categories in which objects can be placed, namely the transient and the durable (Thompson, 1994). As Jonathan Culler explains,

> Objects placed in the transient class are thought of as having finite life-spans and as decreasing in value over time. Objects viewed as durable are endowed with, ideally, infinite (but at least very long) life spans, and retain their value or even increase in value over time. (1988, p. 171)

Typically, transient objects are everyday things such as clothes, tools, and food, whereas durables are works of art, monuments, and religious objects. As a norm, transient objects, when they are not consumed entirely or destroyed with use, turn into rubbish or junk. It is possible for durables to turn into transients (such as in the case of statues of Lenin in the former Soviet Union) and vice versa (as in the case of antiques or nostalgic collections); the intermediary stage is their transformation into junk. Thompson uses "junk" or "rubbish" to refer to those things that people accumulate in their attics, but we think that "dispensable objects" is a more accurate term. When an object turns "dispensable," it can be either elevated as a durable or demoted as waste. This junk phase seem to correspond roughly to what Victor Turner has defined a cultural stage of "liminality." Liminal entities are outside of established structures of meaning, accumulated at random in the attic of culture, as it were. This stage of liminality corresponds to a *tabula rasa*, in which the previous value and status of objects are forgotten, denied, or annulled. Given that any object—even a handful of dirt—can be isolated and become meaningful, the junk phase is temporary, even though it can last indefinitely. Social transformation often involves a redefinition of objects, which go through a liminal stage of "disposability." (On liminality, see Turner, 1995.) This liminal state is crucial, because it determines the future destiny of the object: it could be either definitively destroyed (turned into waste, or into its brute materiality) or abandoned, and therefore lose its cultural meaning, or it could be re-promoted—in this case, with a different status, be it sacred (as a religious object) or profane (as an art object).

Of course, such a liminal phase in the status of sacred objects is a heuristic device, a useful category for thinking about the life span of objects, and its definition depends on the observer's point of view. In fact, a junk phase is probably neither universal nor even necessary. In most cases there are always competing views of objects, and sometimes the people who hold a hostile view get the advantage. If an icon is sold off to a collector, there might be no junk phase, just some locals who revere the icon, a local enabler who just sees it as something to steal and sell, and the collector who values the object for his own reasons. In such cases there is no liminal phase, because no one's view has changed, and the object was never valueless.

However, as in Victor Turner's theory of life cycle/life crisis theory of human beings and society, the idea of a liminal stage in the status of an object becomes relevant when we consider the "life" of that object as an overall process with successive investment of values, rather than in terms of actions that human beings do to that object and their motivations. (Also for a human being, perhaps, there are no such thing as rites of passage; one day you do this, the next day you are asked to do another thing, and then you do still another.) A liminal phase is particularly helpful to explain cases of humiliation and defacement, which are not really necessary if the iconoclasts simply want to get rid of, say, a silly relic from a superstitious past or to recycle those objects. Humiliation and defacement presuppose a previously exalted state, which they aim to negate, and this reversal necessary implies a liminal state in which the passage between the exalted and the debased takes place.[10]

In the case of buddha images, raw materials (which are transient) are turned into durable objects—doctrinally speaking, the most durable objects: Buddha-bodies, which (when considered in their absolute condition as Dharmakāya) continue to exist even without visible form, but yet are visible everywhere on temple altars. The union of the visible and invisible is carried out by multiple ritual actions and empowerment, as we have seen in Chapter 1. This kind of magic is connected with what is perhaps one of the central issues in the study of religion: the transformation of the profane (raw materials) into the sacred (a Buddhist icon) or, in other words, the creation of the sacred as a meaningful category. For example, some Buddhist traditions, especially in Japan, developed a theory of the sacred in which the raw materials are always/already sacred. Doctrines of the Buddhahood of non-sentients, usually known in Japan as "plants and trees become buddhas" (*sōmoku jōbutsu*) were employed to emphasize the intrinsic Buddha-nature of icons—which therefore are Buddhas not just because they are the inanimate containers of the "spirit" of a Buddha, but because their own material is already imbued with Buddha-essence (Rambelli, 2007, pp. 143–4); this is expressed by the idea of taking out the Buddha image already present in the wood (ibid.). The difference between the wood of the icon and the wood of the woodchips on the workshop floor is only that the icon's Buddha-nature is now manifested in terms we can perceive better. The icon is the material world communicating Buddha to us.

In practice, most cases are more complicated. Origin narratives sometimes report that icons enshrined today as secret Buddhas (*hibutsu*) were, in previous times, "normal" icons, exposed to private or public worship, that had subsequently been subjected to acts of iconoclasm broadly understood: defaced, damaged, abandoned. This is the case of the famous Zenkōji icon, the Amida triad, supposed to be the first Buddhist icon to arrive in Japan in 552 and almost immediately becoming the object of anti-Buddhist persecutions; rescued, it was later enshrined at Zenkōji. (On this icon, see MacCallum, 1994.) The Sensōji Kannon in Asakusa, Tokyo, was found floating in the sea by a fisherman and enshrined in a temple as a visible icon; it became secret only following a revelatory dream in which the image of Kannon appeared to an itinerant priest and instructed him to enshrine the statue as a *hibutsu*.

The Yumedono Guze Kannon at Hōryūji went through a dramatic passage from premodernity, as a secret icon, and modernity, as a "national treasure" (*kokuhō*) artwork *qua* Buddhist icon. Significantly, this transformation was marked, at least in the accounts provided by its two modern "discoverers," by a passage through a state of "abjection" (Kristeva, 1982). The beautiful statue had become a nest of spiders, rats, and snakes. Okakura Tenshin wrote:

> The mustiness accumulated over a thousand years almost overcame us. When we cleared away the spider webs we saw a desk, which is probably from the Higashiyama period; and beyond it, we could touch the statue. It was 7 or 8 feet tall and wrapped endlessly with cloth and pieces of sutra. We were startled when snakes and rats suddenly appeared; perhaps it was the surprise of signs of life. (1939, vol. 4, pp. 54–5; quoted in Tanaka, 1994, p. 28)

Ernest Fenollosa also reports that "our eyes and nostrils were in danger of being choked with the pungent dust" (1911, vol. 1, pp. 50–1; quoted in Tanaka, 1994, p. 24). As this account makes clear, the transformation of a kind of durable (sacred icon) into another one (art object) goes through a liminal "rubbish-like" stage of disposability, as indicated also by the events of anti-Buddhist persecutions in the early Meiji period, in which sacred objects were treated as "rubbish" in Japan and as precious objects abroad.

To sum up, changes in the semiotic status of a durable object go through transitional states of disposability, in which the object in question is culturally demoted, treated as substance rather than significance, and then given new meaning according to two possibilities: it is either sold off or burned or neglected, or it is rediscovered and revalorized. The same seems to apply also to bodies. During anti-Buddhist persecutions the symbols of monkhood were targeted (often, together with the actual persons/monks); with disrobing, former monks were reduced to a liminal condition as no-longer-monks-but-not-yet-ordinary-subjects; after that, they could either be re-integrated within society a commoners or, at times, regain their religious status (as either monks or, as in early Meiji Japan, as Shinto priests).

For a semiotics of destruction: Semioclasm and hieroclasm

Changes in the meanings of objects are not simple, but can be perceived in multiple ways, and go through multiple stages. We have argued that the display of religious objects in a museum can be a form of iconoclasm, in the sense that it alters the semiotic structure of the object displayed, in some cases by stripping it of its previous sacred value and turning it into something else, such as art object, cultural property, institutional symbol, pop icon, or propaganda mechanism. Still, the "objects themselves" remain intact and even well-cared-for. Conversely, the removal of an icon from any gaze was a technique that generated increased sacrality in certain objects. In other words, iconoclasm is not a discourse or a set of practices affecting only the physical of sacred objects, but also and, we would emphasize, *especially* their semiotic structure. Thus, the complexities of iconoclasm require a concept of *semioclasm*.[11] Semioclasm is an important cultural activity, perhaps coextensive with the cultural field of destruction. Now, in semiotic perspective, destruction, and iconoclasm in particular, targets material entities as the bodies of specific sign-objects and, more importantly, their semantic fields. Iconoclastic acts, thus, target not just objects, but signifiers and, more crucially, their meanings, the relations between signifier and signified, and more or less extensive portions of the semiotic system underlying the existence of the sign-object.[12] In this perspective, if a sacred object is completely destroyed but replaced with another one with the same function, this instance of iconoclasm may have more limited effects on the underlying cultural system than when a sacred object is materially preserved but radically redefined.

In order to better understand the mechanisms and effects of semioclasm (the destruction of meaning), it is necessary to address the semiotic system and the modes of signification traditionally associated with sacred objects targeted by violent actions. For the purpose of this book, we will follow Umberto Eco's semiotics of architecture (Eco, 1980, pp. 189–249), and adapt it to our discussion of Buddhist objects and bodies. According to Eco, each architectural structure is a signifier connected with a complex signified. On the primary level of denotation, the signified of a temple is the set of its functions, clearly established and culturally defined—in short, a mode of worship and interaction with the sacred. On a secondary level, a temple refers to a set of "symbolic" functions such as general conceptions of religious practice, ideas about cosmology, representations of the sacred, and forms of power; in other words, a temple connotes the "overall ideology" that has produced it and controls its practical and symbolic functions. It is important to notice that primary functions can exist only in presence of symbolic functions. As Eco explains, "'symbolic' connotations of an object . . . are no less 'useful' than its 'functional' denotations" (ibid., p. 206). In some cases, symbolic functions can be more important than practical functions: a temple can be destroyed not because of the specific modes of worship that takes place in it, but because of the symbolism associated with it. All these functions are intersubjective, culturally determined, and codified (or codifiable in principle).

Expanding on Eco's theory, we propose to distinguish further among the following orders of significance. On the plane of denotation, a temple as a cultural artifact is a sign of its function; as such, it signifies what people normally do in a temple in a given culture and historical period. In addition, there are a number of connotative planes:

(a) theological meaning (doctrinal status of the temple; nature of the representation of the sacred in the temple; salvation to be attained through practice in temples, etc.);
(b) cosmological meaning (relation between the temple and the structure of the universe);

(c) ideological meaning (a temple means and produces prosperity of Buddhism, virtue of government, cosmic harmony, social peace, all defined according to the school, the historical period, etc.).

Umberto Eco is well aware that architectural meanings change throughout history, and proposes six examples of semiotic transformations involving permanence and loss of old functions and acquisitions of new ones (ibid., pp. 210–12). In our specific case of Buddhist iconoclasm, destruction in a broad sense can operate on any (or all) of these levels. Accordingly, the denial of cosmological meaning produces forms of individual and more interior practices; the denial of theological meaning empties a temple of its religious status but preserves it as a cultural artifact (as an example of art history, national identity, etc.). But in certain cases, even the need to preserve such artifacts can be questioned: the very memory of the primary function is to be destroyed, and the temple is reduced to its construction materials (as when a temple is converted into a storehouse or dismantled to recover its wood and metal).

Acts of violence against particularly valuable objects and people often follow precise epistemic and ideological models and thus constitute full-fledged instances of *representations* of cultural systems. In particular, what we would like to emphasize is that iconoclasm is not mere violence against *objects*, but it operates instead essentially on the *signs*, that is, the cultural units (types) of which those objects are particular occurrences (tokens). In other words, the cultural impact of iconoclasm is due to its action on the semiotic systems governing the significance of the objects involved. It is important to distinguish semioclasm from iconoclasm: not all iconoclasm involves semioclasm—as, for instance, in the case of unnoticed or accidental damage. The destruction of an object acts on the semiotic system to a greater or lesser extent, but the destruction of an object does not correspond to the destruction of (attacks on, discrediting of, forgetting of) meanings. The targets of semioclasm may be any of the following items or their combination:

(1) The entire signifier: this amounts to the destruction of the object, even though it does not necessarily affect the latter's meaning. For example, the Great Buddhas of Bamiyan did not cease to function (semiotically, at least) as buddhas after their destruction in 2001. Given that most Westerners had never heard of them until they were destroyed, the meanings of the buddhas actually thrived after the tanks had done their work.
(2) Parts of the signifier (of the object): as in acts of vandalism, defacement, etc.

(3) The code connecting signifier and signified: in this case, for example, religious, comological, and ideological values remain, but they are forced to find other representations.
(4) Parts of the signified (one or more of the semantic levels indicated above): for example, the same object can preserve its religious value, but becomes the expression of a different ideological system; or, it can preserve its ideological value but be transformed into the expression of artistic, and no longer religious, values. Meanings can change in ways that require quite different modes of analysis.
(5) The entire signified: this is a radical negation of the semiotic and cultural status of the object, as when a sacred icon is used in a completely different cultural context as a mere decoration on the wall.

We can see as forms of semioclasm also iconoclastic acts targeting the body of religious specialists. In these cases, signifier and signified fatally tend to coincide with the very body. Thus, if we follow the typology we have just presented, semioclasm can affect the entire body of a monk (the entire signifier), thus causing severe injuries or even death; parts of the body (defacement, injuries, etc.); the code connecting signifier and signified (as in attempts to redefine the status, role and place of the clergy); parts of the signified (as when a picture of a monk is used as a "neutral" description of the Samgha or monkhood without implying any devotional attitude, or when the role and function of monks are redefined); and the entire signified (as in the case of mass disrobing, in which the meaning of monkhood itself is questioned and negated).

From this perspective, a monk's body is not just the body of an individual monk. It is literally a sample of the Samgha as one of its representatives, and therefore one component of the Three Jewels; in this way, one monk comes to embody the totality of Buddhism. The semioticity of the monk's body is not constituted only by its accoutrements (the robe, the bowl, ritual implements, etc.) nor by the tonsure, but also by the body itself: deportment and behavior, as direct manifestations/expressions of the monk's *habitus* (as we have defined it in Chapter 1). This typology of semioclasm generally parallels our catalog of destruction from irreversible damage/destruction to cultural redefinition.

One might think of these distinctions as a range between a "soft" and a "hard" semioclasm. Soft semioclasm operates on the sign of the culturally valuable object (it respects the symbolic and doctrinal system of Buddhism, but not the form in which it is represented—a specific temple or statue, or the behavior of a community); it can be reversible and, in any case, does not alter the total set of significations of its targets. In contrast, "hard" semioclasm is aimed at the *signified* or, more radically, at the entire symbolic system; its goal is the total destruction of the sign as a whole by the elimination of one or more orders of signification listed above. Obviously, these two types are not completely separate, but rather constitute two polarities in a continuum corresponding to our typology of destruction. Hard semioclasm entails forgetting and cultural amnesia. Once the meaning is forgotten, it is harder to recover it because it is much more contextual and ephemeral: statues and temples are generally more durable and stable than thoughts. They can be rebuilt, restored, or reproduced if a conceptual model exists, but memories once forgotten, are irretrievably lost. In the case of specifically religious objects, semioclasm can entail the partial or total displacement, reconfiguring, or negation of sacrality ranging from single objects to the more general place of the sacred in a given culture. We could call these processes hieroclasm (literally, destruction of the sacred).

A special case is represented by code semioclasm, which affects the principles connecting a signifier (in this case, more precisely, a sacred object or the body of a religious specialist—as a signifier representing the class of objects it stands for and signifies) with its signifieds. When the semiotic code is targeted, any number of different iconoclastic outcomes become possible; in fact, by targeting the codes that govern our understanding of religious objects, monks, and the sacred in general, we can determine the meaning of the destructive act itself—whether it was a remodeling or a disfiguring, a manifestation of humility or humiliation, a sacrifice or a malevolent obliteration, a negative or a positive cultural redefinition. In other words, code semioclasm seems to belong to the realm of ideological discourses, in which acts of destruction are explained, rationalized, and justified. As such, only institutions and collective agents can perform it. Most cases of iconoclasm, in contrast, target either signifier or signified, or both. Thus, in order to avoid radical effects of semioclasm, that is, the destruction of a cultural system (which could happen, for example, if religious institutions should acknowledge that the temples' value is just the use-value of the materials of which they are built), institutions affected must try to preserve the code connecting signifier and signified so as to preserve the coherence and raison d'être of their symbolic system. They can do so, first, by interpreting hard semioclasm (the destruction of the symbolic system) as soft semioclasm (the destruction of some signifiers) and, second, by appropriating the latter to reaffirm and strengthen their own symbolic system. This is what Buddhist institutions actually tried to do in Japan to counter massive destruction, as we saw in Chapter 2. In other words, code semioclasm controls the shift between different registers of destruction to determine whether destruction has negative or positive intentions/effects (that is, to determine if an act of destruction is carried out in order to harm or to preserve).

There is an important consideration about the ways in which bodies are affected by acts of iconoclasm qua semioclasm as differently from mere objects. As we have seen, sacred objects that have been targeted by iconoclastic acts can almost always be repaired or restored, or if not they can be replaced by other objects; at times, even fragments of the original that has been destroyed can be worshiped. These restored or replaced objects may not satisfy the requirements of modern canons of cultural or artistic authenticity, but function well within their religious contexts, where they are usually not perceived as replacements but as "originals." In fact, a sacred object ceases to be perceived and used as such only in cases of radical re-signification, as we have seen in Chapter 4, or in cases of radical cultural transformation, often involving the extinction of religious traditions and more general cultural practices.

However, the body of a holy person (including that of religious specialists such as monks), precisely because it belongs to a human being, can never by definition be replaced. This is where we ought to draw the line in rationalizations of iconoclastic acts. We may understand acts of desacration involving religious *symbols* attached to the body of religious specialists (robes, ritual implements, etc.), as expressions of religious, political, or cultural criticism (in some cases, we may even agree with these acts), but we cannot condone physical harm done to human beings; accordingly, it is important to find iconoclastic strategies that target explicitly and exclusively the accoutrements of the body and never the human body.

Conclusion

Destruction and Cultural Systems

Semioclasm is an incessant cultural process. All changes in the semiotic space—the semiosphere (Lotman, 1990)—of a culture (semiomorphosis) require in fact forms of semioclasm, in which signification and communication are reconfigured. It may even be possible to envision a typology of modes of "semioclastic" labor, that is, the actual labor involved in the destruction of semiotic entities, based on a reversal of Umberto Eco's "modes of semiotic production" (Eco, 1975, pp. 151–313). After all, as we have attempted to show, iconoclasm is not a simple and unproblematically blind act of irrational violence, the expression of rage and fury. People do not just pick up a statue and break it into pieces. They have to choose that particular statue (or that set of statues) on the basis of some criteria. Then, they have to decide what kind of destruction they will enforce: total material destruction (burn to ashes?), recycling (statues→bronze→cannons), re-coding (statue→art object; temple→school). They also have the option of selling the statue to collectors (statue→money); or they can just content themselves with some form of carnivalistic and obscene defacement to represent a moment of suspension of social order. What is common to all this, however, is the focus on semiotic processes.

Although there are certain intensified periods of destruction, iconoclasm is best considered as a permanent feature of cultural life. Perhaps relevant here is the idea of "inner iconoclasm" as employed by Christian Kordt Høbjerg to define "people's ability to reflect upon and sometimes even question their own religious beliefs as these are generated and unfold in the course of ritual performance" (Høbjerg, 2002, p. 58). For Høbjerg, iconoclasm is "built into" iconists' ideas and practices. He writes: "Aniconic attitudes, doubt and skepticism represent a reflexive stance that religious agents take toward conventional representations and practices associated with an otherwise invisible order of supernatural agencies" (ibid., p. 64).

The examples we have presented in Chapter 4 point to the fact that even museum display—a form of iconoclasm (or hieroclasm) because it destroys the semiotic mechanisms associated with the religious objects to turn them into something else—can also result in, or be related to, forms of re-sacralization, that is, instances of semiomorphosis in which some kinds of sacred ideas or principles keep being restructured. It would seem that some people cannot discard sacred objects once they have exhausted their function, but try to prolong their lives by giving them a new, semi-religious status. What is interesting to note is that hieroclasm (the destruction of the sacred components of an object) is not a definitive action. Sacrality remains as a trace of a previous sacred state, as a sense effect (as when a statue is displayed as an example of religious art from a certain culture and a certain historical period), in a modified form (as the bodhisattva Myōken now housed at an amusement park), or even as a consequence of resacralization (as when people bow and make offerings to a statue in a museum). In some extreme cases, even the mere signs of the absence of an image that has been destroyed can signify that object, as the disquieting empty caves where the great Buddhas of Bamiyan used to be. In this case, the destroyed object continues to exist in an immaterial form, in pictures, tales, studies, etc. Another instance of resacralization, perhaps more subtle and problematic, occurs when a previously religious object is redefined as a "national treasure" embodying the artistic sense and spirituality of a nation, as we have seen in the previous chapter.

The power that states and important cultural institutions possess to define the semiotic status of an object (sacred? artistic? junk? rubbish?), is based on their power of coercion, or force (i.e. the possibility to legally threaten, and actually exercise, physical violence)—a violence that, as we have indicated, is constantly exercised on signs and signification. We would like to suggest that what keeps together coercion and consent to allow for the reproduction of a hegemonic formation is precisely a systematic semioclasm as a policing of signification. Semioclasm, however, also plays an important oppositional function as the constant possibility of dismantling official, authorized semiotic systems, an activity that allows for the production of new meanings. This intrinsic ambivalence of semioclasm is what makes iconoclastic acts essentially ambiguous. Iconoclasm can be the result of an official policing of signification, and as such can be used by anti-establishment interpreters to point to the arbitrariness, limitations, and coercive nature of official semiotics. On the other hand, revolutionary iconoclasm, which shows the arbitrariness and ugly aspects of official semiotics, can be employed by the threatened, official establishment as an indication of the need to police signification. A particularly powerful case of contested iconoclasm is represented by the revolutionary exhumations in Spain during the Civil War (1931–9), studied by Bruce Lincoln (1989, pp. 103–27). Lincoln writes,

> in some measure society is constructed (and continually reconstructed) through the exercise of symbolic discourse, then the destruction of widely recognized and even revered symbols may be seen as an attempt to undo their effects, that is, to deconstruct the social forms that others have constructed and maintain through them. The iconoclastic act is thus less a matter of the icon themselves that it is an attack launched by members of one segment of society against those of another, designed to isolate the latter and to deprive them of an important instrument with which they have in the past maintained social forms in which they hold privileged positions. (ibid., pp. 117–18)

But here lies a paradox. The attack on a segment of society and, we would add, its hegemonic role, is almost always associated with an attack on particular objects and/or bodies that are widely recognized as its symbols. As a consequence, those objects become more than mere signifiers of a semiotic and ideological system, since iconoclasts charge them with a particular "spirit" that they come to embody. It is not necessarily true, then, that, as Bruce Lincoln writes,

> no act of iconoclasm is ever carried out with the intent of destroying an icon's power, for iconoclasts . . . act with the assurance either that the specific image under attack has no such power or the more radical convinction that there is no such thing as sacred power. (ibid., p. 120)

On the contrary, the power of iconoclasm lies precisely in the fact that it targets a specific "spiritual entity" that is supposed to permeate the object, and as such it is probably not different from a radical instance of fetishism, in which a sacred object is abused when it does not work and is finally discarded when no longer needed.

The subject of religious violence is definitely "disturbing" (Davis, 1975, p. 185); perhaps it is for this reason that it is explained away as "an extraordinary event, the product of frenzy, of the frustrated and paranoiac primitive mind of the people" (ibid.) or perhaps as a temporary mistake. However, it is probably more productive to assume, again with Natalie Zemon Davis, that "conflict is perennial in social life, though the forms and strength of the accompanying violence vary, and that religious violence is intense because it connects intimately with the fundamental values and self-definition of a community" (ibid., p. 186). Religious violence is the product of patterns of behaviors, systems of meaning, and ideologies in a given culture at a given historical juncture. The study of religious destruction is therefore a particular instance of what Michel de Certeau called a "*polemological* analysis of culture." In de Certeau's words,

> culture articulates conflicts and alternatively legitimizes, displaces, or controls the superior force. It develops in an atmosphere of tensions, and often of violence, for which it provides symbolic balances, contracts of compatibility and compromises, all more or less temporary. (de Certeau, 1984, p. xvii)

This sheds light on a famous line by Walter Benjamin: "The tradition of the oppressed teaches us that 'the state of emergency' in which we live is not the exception but the rule" (1968, p. 257).

The cases we have discussed in this book, chosen because they are representative samples of larger, deeper, and widespread cultural phenomena, point to the presence of violence not only on the level of bodies and matter in human societies, but, more fundamentally, as a way to interact with semiotic systems. This is the essential and important difference between our idea of social violence and other, more generally accepted visions such as those by René Girard (1977) and Walter Burkert (1983) (see also Hamerton-Kelly, 1987). These authors, each in his own way, end up by reducing violence to an ontological dimension, and therefore fail to recognize its function in the creation and maintenance of concrete social formations. In our view, instead, every transformation in the regime of symbolic systems requires some acts of violence, either on the signifiers or, more effectively, on the signifieds. Semioclasm thus appears to be one of the fundamental mechanisms for the creation and preservation of social and cultural orders as the intersection of what Antonio Gramsci called coercion and consent, the two fundamental components of every form of hegemony. Consent—or, as Bruce Lincoln calls it, "discourse" (and it is superfluous to indicate the Foucauldian connotations of this usage)—is essential to create and transform the consciousness of the members of a given social formation, so that they can come to accept it and participate in it. (In this case, acceptance includes also injustice and exploitation intrinsic to that social formation.) But consent is the result of a "policing of signification" that works on two levels, namely, it defines concepts, ideas, and statements as acceptable or unacceptable, and it attempts to prevent unacceptable ideas to be produced and circulated. Such a policing of signification is a systematic form of semioclasm that destroys new, unauthorized signs and reconfigures old signs in acceptable forms.

We (rather boldly) posit the possibility of a new disciple of Destruction Studies, which would be far greater than the categories and ruminations of this book. Is it so absurd to imagine universities with departments of Destruction Studies? This academic discipline would not be just devoted to description and analysis, but would be able to develop ethical and political positions about cases of cultural destruction. While acts of destruction can at times be inevitable and even, perhaps, necessary to establish new and more liberational cultural forms and norms, still we need conceptual tools to address (not just intellectually, but also and especially in political and practical terms) acts of violence that threaten people and culture. Having seen how mutually implicated destruction and production are, how can we escape the dichotomy that nonetheless remains? A clue for better envisioning the contours of this discipline can perhaps be found in Roland Barthes's concept of "subtle subversion." Barthes wrote,

> By *subtle subversion* I mean [. . .] what is not directly concerned with destruction, evades the paradigm, and seeks some *other* term: a third term, which is not, however, a synthesizing term but an eccentric, extraordinary term. An example? Perhaps Bataille, who eludes the idealist term by an *unexpected* materialism [. . .]; thus Bataille does not counter modesty with sexual freedom but . . . with *laughter*. (1982, p. 409)

In the same way, destruction studies as we envisioned them, as a polemological study of culture, should be able to identify and propose subtle, nonviolent subversions that, while contributing to cultural change and innovation aimed at human liberation, neither harm people nor important elements of the cultural landscape.

Notes

Chapter 1

1 A measure of the range of objects can be gained from dictionaries of Buddhist sacred implements (Jp. *butsugu*, Ch. *foju*), such as Okazaki (ed.), 1982; Shimizu (ed.), 1999. A similar publication in Chinese is Zhulin, 1992. Among premodern sources, the *Sanbō butsugu shō* ("Catalogue of Objects and Ritual Implements of the Three Treasures"), is an illustrated collection of a large number of tools and objects that defined the life of a high priest in a major temple, and constitutes a sort of chart of the Buddhist system of objects in the Edo period; in Takakusu and Watanabe (eds), 1924–35, vol. 10, pp. 1269–380.

2 Within the Sōtō Zen tradition, *kirigami* documents focus on the Dharma robe as equivalent to dynastic treasure. On Dharma transmission in Sōtō Zen *kirigami*, see Faure, 1996, pp. 57–62; on transfer of material objects and wealth within the Tendai tradition, see Stone, 1999, pp. 138–48.

3 These expressions referring to the nonsentients and the realm of objects are related to concepts such as "sentient beings" (Jp. *ujō*, Ch. *youqing*), "living beings" (*shujō*, *zhongsheng*), "realm of sentient beings" (*shujōkai*, *zhongshengjie*), and "realm of the Buddhas" (*bukkai*, *fojie*). The karmic environment (*ehō*, *yibao*) is related to "karmic retribution proper" (*shōbō*, *shengbao*), the particular body–mind complex that forms the subjectivity of a sentient being as a result of karma.

4 Other, more ancient factors need to be considered, such as the beginning of the cult to the Buddha, the development of cults to cosmic buddhas different and transcending Śākyamuni like Amida, Miroku, Kannon, Yakushi, etc.

5 See for instance, *Zeng yi ahan jing* (Sk. *Ekottarāgama*), in Takakusu and Watanabe (eds), 1924–32, vol. 2, no. 125, p. 699a; *Foshuo zhude futian jing*, in ibid., vol. 16, no. 683, p. 777b.

6 Significant exceptions are John Kieschnick, with his thorough description of merit-making activities involving public works (Kieschnick, 2003, esp. pp. 199–214), and Janet Goodwin (1994), with her emphasis on construction works as important vehicles for the diffusion of Buddhism and the local rooting of religious institutions. More recently, Amy McNair (2007) has analyzed the complex interactions of religion, politics, and patronage in the construction of Longmen monuments in China.

7 See, for example, Faure, 1998, Sharf, 1999, Ruppert, 2000, and Kieschnick, 2003. Note also studies of Chinese monastic economy, such as Twitchett, 1955–6, Gernet, 1995, Prip-Moller, 1991, and Walsh, 2010.

8 Daoxuan, *Shimen guijingyi*, in Takakusu and Watanabe (eds), 1924–32, vol. 45, no. 1896, p. 856.

9 Yanzong, *Jishamen buying baisu dengshi*, in ibid., vol 52, no. 2108, p. 456.

10 Daoxuan, *Shifenlu shanfan buque xingshichao*, in ibid., vol. 40, no. 1804, p. 147.

11 Bourdieu stresses the non-linguistic nature of *habitus*: "the factors which are most influential in the formation of the *habitus* are transmitted without passing through language and consciousness, but through suggestions inscribed in the most apparently insignificant aspects of the things, situations and practices of everyday life" such as "modalities of practices, the ways of looking, sitting, standing, keeping silent, or even of speaking" (Bourdieu, 1991, p. 51).

12 Yanzong, *Jishamen buying baisu dengshi*, in Takakusu and Watanabe (eds), 1924–32, vol. 52, no. 2108, pp. 461c–462a.

13 Daoxuan, *Shimen guijingyi*, in ibid., vol. 45, no. 1896, p. 862.

14 Yanzong, *Jishamen buying baisu dengshi*, in ibid., vol 52, no. 2108, p. 855.

15 *Shimen guijingyi*, in ibid., vol. 45, no. 1896, p. 858.

16 Bernard Faure mentions a Sri Lankan ritual of "closing the eyes" "to deconsecrate it or make it temporarily powerless": Faure, 1996, p. 249.

17 Letter from Miss. M[ary] Darley, in a composite article, "Our Twenty-Sixth Anniversary Meeting," *India's Women & China's Daughters*, 1906, p. 86.

18 *Shimen guijingyi*, in Takakusu and Watanabe (eds), 1924–32, vol. 45, no. 1896, p. 857.

19 Ibid., p. 858.

20 *Da baoji jing*, in ibid., vol. 11, no. 310, pp. 507–14. See also *Foshuo fa miejin jing*, in ibid., vol. 12, no. 396, p. 1119a.

21 *Mohe Moye jing* (Sutra of Mahāmāyā), in ibid., vol. 12, no. 383, p. 1005, 1013c–1014a. Several texts in the Buddhist Canon describe the end of the Dharma with apocalyptic visions of destruction, in particular the *Fa miejin jing* (Sutra of the Annihilation of the Dharma), in ibid., vol. 12, no. 396.

22 Nattier (1991) discusses various timetables, including periods of 500, or 1,000 years.

23 In Takakusu and Watanabe (eds), 1924–32, vol. 45, no. 1897; the quotes are, respectively, on pages 870a, 870a, 870a, 870b, 871b, 870c, 870c, and 870c.

24 Ibid., vol. 16, no. 696, pp. 797c–98c. On this subject, see also Boucher, trans., 1995.

25 In Takakusu and Watanabe (eds), 1924–32, vol. 16, no. 698, p. 800b.

26 Kakuban, *Kyōyōshū*, in Bussho kankōkai (ed.), 1978, vol. 43, pp. 4a–b.

27 For instance, the Ashikaga Shoguns were instrumental in building in many areas of Japan temples called Ankokuji (lit., "temple for the pacified country"), in order to pacify the spirits of the people who died during the fifteenth century internal warfare.

28 It is defined as "a movement against the worship of holy pictures in the Eastern Church, in the eighth and ninth centuries, which was repeated on a smaller scale in the Frankish kingdom": Fortesque, 1913–22, p. 78.

29 Virgil Candea writes that the term "refers to opposition to the veneration of images. It was a trend of thought that caused great disturbances in the religious life of

the Byzantine empire . . . and engendered debates that led to the clarification and elaboration of the theology of icons": Candea, 1987, p. 1.

30 The similarity of these aspects between European and East Asian iconoclasms has already been indicated by Paul Demiéville (1974).

31 The above is based on Besançon, 2000, p. 114.

32 "The government was specially fierce against the monks, as being the chief defenders of images. Monasteries were destroyed, monks banished, tortured, and put to death": Fortesque, 1913–22, p. 78.

33 In 727, Emperor Leo III (r. 717–41) issued an order that the image of Christ over his Palace gate should be destroyed. Runciman describes the result: "There was a popular riot; and the official who was carrying out the Emperor's instructions was done to death by furious women. The rioters were severely punished, by scourging, mutilation or exile. This provoked further unrest. Leo could count on the support of the army" (1977, p. 68). Note that it was women who were the angriest about the iconoclasm; also note that the image was, legally, the private property of the Emperor himself—he was destroying his own property. In a similar case later, Emperor Leo V (r. 813–20) emphasized his sponsorship of iconoclastic polemics "by encouraging soldiers to mutilate the image of Christ over the Palace gate" (ibid., p. 85).

34 The Patriarch of Constantinople (from 730 to 753) Anastasius at age 70, was humiliated by Constantine V (r. 741–75), when the latter recaptured the capital from his brother-in-law Artavasdus: "Anastasius was paraded in the Emperor's triumphal procession, seated on an ass with his face to its tail, being scourged as he rode along. But he remained on the Patriarchal throne" (ibid., p. 71).

35 In 761, Constantine V's persecution of iconodules (image worshipers) focused largely on monks. "Monasteries and nunneries were pulled down and their inhabitants ordered to marry or else be exiled. In 765 a number of recalcitrant monks were forced to parade in the Hippodrome, each escorted by a noted prostitute. At least six monks were put to death [. . .] certain provincial governors won the Emperor's approval by their exceptional savagery" (ibid., p. 75).

36 Some of these themes have been discussed in a powerful and impressive way in the novel *Q* by the collective of Italian writers known as Luther Blissett (2005).

37 Besançon also points to the existence of iconoclastic tendencies at the basis of modern Western art, especially in its symbolist and avant-garde movements, in which the possibility of figurative representation of reality is denied in favor of abstraction transcending objects and figurative imagery. In Besançon's words, that of the new artistic movements "was a new iconoclasm, given that their renunciation of reference to 'objects' and to nature did not stem from a fear of the divine but from a mystical goal of providing an image that was finally worthy of it" (Besançon, 2000, p. 7). Iconoclasm can thus paradoxically express a quest for the ultimate image.

Chapter 2

1 I owe this story to Fujita Ryūjō, personal communication, June 2001.

2 The physical grounding in space and the material aspect of sacred places have been particularly emphasized by Mircea Eliade (1959). More recently, and with different approaches, Jonathan Z. Smith (1978, 1987) and, in the specific case of Japan, Allan Grapard (1982, 1994b, 1998) have shed new light on these issues.

3 On Rennyo's iconoclasm, see Blum, 2006 and Chiba, 1990.

4 For a general view of this subject, see the series *Ikki*, in particular Aoki et al. (eds), 1981; see also Katsumata, 1982.

5 Jonathan Z. Smith has pointed out that some radical religious organizations promote a psychology of self-destruction as a way to provoke or accelerate the inevitable apocalypse—and their own apotheosis (1982, pp. 102–20). In cases of extreme danger for the survival of important temples, their monks may have developed a similar psychological attitude—or, at least, acted as if they did. No one in medieval Japan was interested in speeding up the final extinction of the Dharma, because that would have meant not only the end of Buddhism, but the collapse of the entire universe.

6 The *Lotus Sutra* states: "Suppose someone should conceive a wish to harm you, should push you into a great pit of fire. Think of the power of that Perceiver of Sounds [the bodhisattva Kannon] and the pit of fire will turn into a pond!"; Hurvitz (trans.), 1976, pp. 303–4.

7 For further information on the issues in this section, see McMullin, 1984; Tsuji, 1955, vol. 10, pp. 1–404; Oishio, 1988; Aneo, 1988; Gonoi, 1990.

8 The sheer amount of violence and destruction that took place is indicated also by the strange, haunted atmosphere that lingered on Mt. Hiei after its destruction. The cloistered prince Gyōnyo, for three times the abbot of Enryakuji between 1663 and 1693, wrote in his diary a series of strange, supernatural phenomena involving the ghosts of the people that had been killed more than 100 years before. See Murayama, 1990, pp. 375–80.

9 A representative example of this tendency is the Zen priest Suzuki Shōsan (1579–1655); see Tyler (trans.), 1977.

10 On anti-Buddhist policies in Korea, see, among others, Goulde, 1985; deBary and Haboush (eds), 1985; and Deuchler, 1992.

11 For a detailed discussion of anti-Ikkōshū policies in southern Kyushu, see Saeki, 2003, pp. 95–114. According to Saeki, repressive policies of the Satsuma domain alone may have caused up to 140,000 casualties; ibid., p. 114.

12 On Meiji period iconoclasm, see, among others, Ketelaar, 1990; Yasumaru, 1979; Tamamuro, 1977; Tsuji, 1984, vol. 4, pp. 142–242; Tamura, 1988, pp. 69–87. Specific case studies of "separation" include: Grapard, 1984, 1992; Antoni, 1995; Inoue, 2003; Liscutin, 2000; Thal, 2002.

13 I owe this information to a conversation with the guide at the shrine's museum.
14 It is significant, though, that the Shinto deities most diffused and worshiped today in Japan are Inari, Hachiman, and Tenjin, that is, some of the most popular syncretic deities created in Japan by Buddhist priests in the Nara and Heian periods.
15 On these subjects, see also Kashiwahara, 1990; Yoshida, 1970; Victoria, 1997.
16 See, for example, the hyperbolic statements of praise for the Kanshinji Nyorin Kannon by Japanese scholars quoted in Bogel, 2002.
17 On the latter definition, see for instance the *Gukanshō*, a Buddhist history of Japan by the Tendai monk Jien (1155–1225) written in 1219, concerning the destruction of Tōdaiji and Kōfukuji by the Taira army: "'Disgusting' is not a sufficiently strong word to describe what happened" (Brown & Ishida (trans.), 1979, p. 128).
18 Raigō's anger and revengeful attitude toward Enryakuji is recorded in the *Tale of the Heike* (Kitagawa & Tsuchida (trans.), 1975, vol. 1, pp. 177–9), but the story presenting Raigō as a rat appears in the *Taiheiki* (Gotō and Kanada (eds), 1960–2, vol. 2, p. 91); Raigō's theme was also exploited by Edo period writers and illustrators of ghost stories; see Yamamoto, 1998, pp. 2–43.
19 The aristocrat diary *Shigetaka-ki* asked: "Isn't the fight between north [Enryakuji] and south [Kōfukuji] like what happens in the realm of the Asura?"; quoted in Kuroda, 1980, p. 61.
20 Kakuban, "Hongan Shōnin gokajō chūmon," in Miyasaka (ed.), 1989, tome 2, p. 291.
21 See also Taussig's further exploration of the connections between state and violence in Taussig, 1992, pp. 111–40 and Taussig, 1997.

Chapter 3

1 In 1742, the Vatican declared the debate over: such rituals were incompatible with Christianity. Only in 1935 was this policy amended to allow a qualified and explicitly "civil" veneration of Confucius; in 1939 a bowed head in veneration of ancestors was permitted. This revision to long-standing orders against obeisance to Confucius was impelled by the newly established Manchukuo and its ideology of Wangdao (Royal Way) requiring all citizens to bow to Confucius. Church leaders were assured by the Manchukuo regime that the cult was strictly civil. Pope Pius XI allowed a head bow, a decision further endorsed in 1939 by Pius XII. At the same time Pius XII allowed the bowing of the head toward ancestral altars, with a further clarification and consolidation of policy in 1941.
2 Similarly: "It is not more sinful and absurd for a Chinese to make prayers to the spirit of his ancestor, than for a Romanist to pray to his patron saint; nor is the invoking of Confucius a grosser error than the invocation of the Virgin Mary" ("Chinese Missionary Work," 1851, p. 111).

3 The quotation here is from 1 Sam. 15.22, where Samuel rebukes Saul for excessive attention to ritual. Another common text for reading was Acts 17.16–31, Paul's anti-idolatry talk at Athens. The hymn "Jesus Shall Reign" by Isaac Watts describes the global victory of the Gospel, even in "heathen lands."

4 Very similar phrasing is found elsewhere: "he brought out his household gods and ancestral tablets, and set fire to the whole in the presence of his astonished and horror-stricken neighbours." ("Cases of Conversion to Christianity at the Cities of Lo-nguong in the Provinces of Fuh-chau, China," 1867, p. 81) For other examples of performances of repudiating gods in the presence of others, see Lutz and Lutz, 1998, p. 18.

5 Another example: A Captain Li Ming converts and sets about destroying his household gods. His three wives and neighbors are alarmed. "He tore all the gods down and brought them into the center of the court. Opening the catechism where it said, 'Thou shalt have no other gods before me,' he applied a match to the pile and the gods went up in smoke" (Goforth, 1931, p. 71). Onlookers predicted calamity, which never happens. There is also a story about the Catholics of Peking who, before accepting new converts, had them bring their Buddhist images to the church, destroy them, and throw them into the church pond. One day, however, just as they were beginning to melt down the images, there was a sudden clap of thunder and all of the idols vanished. The people who had assembled broke into a sweat, clasped their palms together, and chanted "A-mi-t'o-Fo" (Cohen, 1963, p. 37).

6 Another example of this tactic of covering doors and floors with "filth" (presumably excrement), for example, Phillips 1893, p. 3. Phillips goes on to describe a mob burning down their house, and beating up Chinese Christians. The use of feces as a symbol of violent transgression is very evocative and probably universal. Erving Goffman describes forms of "negative deference" (Goffman, 1967, p. 88), such as thumbing one's nose, etc. An extreme form seen in psychiatric wards: "Whatever is in the patient's mind, the throwing of feces at an attendant is a use of our ceremonial idiom that is as exquisite in its way as is a bow from the waist done with grace and flourish" (Goffman, 1967, p. 89).

7 Other voices suggested more mundane reasons. A. E. Moule proposed several possible sources of such violence: "Some will trace in the violent and savage scenes at Wu-sieh and I-chang retribution for the anti-Chinese mob violence at San Francisco, and for the relentless anti-Chinese policy pursued in Australia" (Moule, 1891, p. 163).

8 Psalm 115, v. 4–8. (Spence, 1996, p. 97) However, in Boardman's evaluation, the Biblical influence was still slight: "The Taipings were undoubtedly happy that the Old Testament took a position similar to the one in which they found themselves on the worship of other gods and images, but they had no occasion to borrow the arguments of Hebrew leaders" (Boardman, 1952, p. 66).

9 Also: "There is no means of computing how many religious edifices we passed on our way to Lanchow that were falling or had fallen into decay, that had been abandoned entirely except for a beggar or two posing as priests, or had become noisome dens in which thieves divide their booty and vagrants scatter their filth; that the traveler may see in almost any part of China" (Franck, 1923, p. 478).

10 At the same conference, Dr Douglas comments: "Napoleon used to conquer by concentrating his troops at one point. Pour shot and shell into the most assailable part of the earthwork; that taken, we shall be able to follow up our success" (Yates et al., 1879, p. 113).

11 Physical vigor and spiritual vigor were inextricable: As missionary R. H. Graves wrote in his speech titled "How shall the Native Church be Stimulated To more Aggressive Christian Work," "faith partakes in the *general quickening* of the faculties excited and developed by aggressive work. In our bodies the quickening of one function involves the quickening of others—if the flow of the blood be quickened by vigorous exercise, the appetite is improved, the power of digestion is increased, the limbs grow strong and the brain is clear. So it is in our souls" (ibid., pp. 341–2).

12 "The temples were not torn down and the idols were not destroyed. Generally the idols were put in the background and veiled. The ignorant but devoted women saw to that" (Woodbridge, 1919, p. 114). Also, the "boyishness and fishwife credulity with which these men solemnly carried out their superstitious antics would have seemed even more out of place but for their girlish cues and their generally simple, almost childlike manners" (Franck, 1923, p. 277).

13 Feuchtwang quotes a vice-governor of Fujian, listing: "witchcraft, sorcery, use of elixers [sic], fortune-telling, astrological practices, invocation to avert calamities, rain-making, supplication for offspring, treating disease with exorcism, practice of physiognomy, practice of geomancy, building village temples" (Feuchtwang, 1989, p. 47).

14 *Nagel's Encyclopedia Guide China* for 1968 reports that the Dongyue miao is "now used as a school," but it is clear from the description (of statuary on the porticos) that the walls surrounding the buildings were not in place at the time this text was written. The brief note mentions a "terrace with a little pavilion" which does not exist today (*Nagel's*, 1968, p. 942).

15 The least affected religion was probably Islam, due in part to a special policy against antagonizing the Hui, a large and mostly Muslim ethnic minority.

16 In Jun Jing's account of forced relocation, villagers were unable to remove their ancestors' remains systematically, and many tombs were hauled away in trucks by construction workers. These remains were described as part of the "twenty tons of garbage, human waste, and material detrimental to people's health" (Jing, 1996, p. 80). The bones that were removed, in used cement bags, got jumbled together.

17 As Yuan remarks, one problem for Red Guards was the intangibility of the targets: "We looked around for something tangible to overthrow" (Yuan, 1987, p. 92).

Chapter 4

1 Particularly innovative in this respect is Singapore's Asian Civilisations Museum, with its accurate descriptions (which include broader background information) of objects on display, interactive devices, and special spaces simulating original settings of objects to enhance a sense of realistic experience.

2 See James Hevia's use of the term "centering" to describe Chinese constructions of the process of approaching the imperial throne (Hevia, 1995, p. 123).

3 Belcher continues: "Once the most humble of utilitarian objects is afforded the treatment of being framed within a glazed space and illuminated (as it must be if it is to be protected yet visible), it immediately takes on the preciousness of an art object and the public may find it difficult to see it an any other way" (1991, p. 147).

4 This episode is dramatically described in the fourteenth-century epic *Tale of the Heike* (Kitagawa & Tsuchida (trans.), 1975, vol. 2, pp. 676–7).

5 We owe this information to Mitsuhashi Tadashi.

6 Again, there are of course exceptions, as in the case of badly damages images (either because of persecutions or because of accidental fire) on display in some peripheral areas of Buddhist temples in Japan.

7 However, it appears that during the Tokugawa period this image was considered to be the *honjibutsu* (the original buddha) of Sannō Gongen, the Tendai tutelary deity of the Chūsonji compound—an icon of *honji suijaku* combinatory religiosity. See Sasaki, 1999, pp. 168–9.

8 The above account of the modern vicissitudes of this image is based on an interview with an anonymous temple guide in October 2000. For additional information on this *hibutsu*, see Yiengpruksawan, 1998, pp. 179–84 *passim*, 193–8 *passim*; Sasaki, 1999, pp. 162–71.

9 For more detailed discussions of *hibutsu*, see Rambelli, 2002; Horton, 2007, pp. 156–92.

10 An additional interesting phenomenon in this respect is the proliferation of websites advertising dates of display and the virtues of each secret Buddha.

11 On the Seiryōji image and its legend, see Henderson and Hurvitz, 1956. This Japanese narrative is based on a number of scriptures, such as *Zengyi Ahan jing*, in Takakusu and Watanabe (eds), 1924–32, vol. 2, pp. 706–8; *Tai fangbian fo baoen jing*, in ibid., vol. 3, p. 136b–c; *Guanfo sanmei hai jing*, in ibid., vol. 15, p. 678b; *Dasheng zaoxiang gongde jing*, in ibid., vol. 16, p. 791a–b. For other details on this legend, see Coomaraswamy, 1927, Carter, 1990, Soper, 1959, Sharf (trans.), 1996b, and Swearer, 2004. On the benefits resulting from worship of buddha images, see, for example, Sharf (trans.), 1996b.

12 This account is based on an interview with an anonymous monk at the Sennyūji in November 2004. On Lady Yōkihi in Japan, see Graham, 1998.

13 On this subject, see Takatsuki and Yoshino, 2004; Itoi and Lee, 2005.

14 This is one of the main points of Lopez, 1998.

15 See the park website at www.yomiuriland.co.jp/institution/park.html.

16 In general, "the heart/soul of Japan" (*Nihon no kokoro*) refers to various aspects of Japanese language, culture, and art, as is clear from, among others, homonymous series of the illustrated magazine *Taiyō* (published in Tokyo by Heibonsha) and of books on aspects of Japanese culture published in the early 1970s by Mainichi Shinbunsha and in the early 1990s by Maruzen. See in particular the NHK (Japan's public broadcast network) documentary series (with explicative books) by Katō Shūichi (Katō (ed.), 1987–8).

Chapter 5

1 Michalski acknowledges that the connection between carnival rites and iconoclasm was originally suggested in a pioneering article by Robert Scribner (1993, p. 94, no. 56).

2 The concept of semiotic labor was first defined by Ferruccio Rossi-Landi (1968), and further developed by Umberto Eco in his theory of semiotic modes of production (1975, pp. 151–313).

3 Partha Mitter presents cases of derogatory attitudes toward Indian images (including Buddhist icons) among European intellectuals; it is likely that display of Indian art in Western museums also served to exemplify such attitudes (Mitter, 1977).

4 For example, the "restoration" of Sainte-Chapelle in Paris during the nineteenth century revealed historical contradictions which made any "authentic" restoration impossible; the "original" was in fact composed over many centuries, and the choice of this or that period as the basic reference provoked debate among architects and political commentators alike.

5 Relevant here is also the practice of violently shaking images in order to activate their power.

6 For example, the so-called father of Tokugawa Neo-Confucianism, Fujiwara Seika (1561–1619) was a Zen monk for some years. Of course there are many lesser-known examples.

7 *Biguan*, "sealed confinement" is "a highly respected religious practice of self-discipline for Chinese monks" (Pittman, 2001, p. 82). For a description, see Fortune, 1857, pp. 186–7, 349, 362.

8 In the eighth- and ninth-century Byzantine iconoclasm, "The government was specially fierce against the monks, as being the chief defenders of images. Monasteries were destroyed, monks banished, tortured, and put to death"

(Fortesque, 1913–22, p. 78). And also: "The monastic habit was forbidden, monasteries were turned into barracks . . . the army especially became fiercely Iconoclast" (ibid., p. 79).

9 For a broader interpretation of the concept of symbolic value as applied to Buddhist material objects, see Rambelli, 2007, pp. 267–9.

10 An example of destruction that took place without a liminal phase might be the Taliban destruction of the Bamiyan Great Buddhas. As far as we can tell, there was no defacement there, just a purely mechanical act of destruction (literally "mechanical," by machine guns, dynamite, and cannon bombing); for the Taliban, the great Buddha was indeed a "thing" without value, a bunch of stones in the desert, and therefore it did not need any act of debasement.

11 Roland Barthes, who first introduced this concept, used it to refer to the effect of critical analysis of received meanings (1972, p. 9), but we propose a broader significance for this term, not limited to intellectual analysis.

12 For a first attempt at developing the concept of semioclasm toward this direction, see Rambelli, 2001.

Bibliography

"A surprise for the gods!" (1919), *Church Missionary Gleaner*. January 12–13.

Abé, Ryūichi. (1999), *The Weaving of Mantra*. New York: Columbia University Press.

Adolphson, Mikael S. (2000), *The Gates of Power: Monks, Courtiers, and Warriors in Premodern Japan*. Honolulu: University of Hawai'i Press.

—. (2007), *The Teeth and Claws of the Buddha: Monastic Warriors and Sōhei in Japanese History*. Honolulu: University of Hawai'i Press.

Ajiki Bun'yū. (2002), *Nijusseki no bukkyō media hakkutsu*. Tokyo: Chōeisha.

Alpers, Svetlana. (1991), "The museum as a way of seeing," in Ivan Karp and Steven D. Lavine (eds), *Exhibiting Cultures: The Poetics and Politics of Museum Display*. Washington, DC: Smithsonian Institution Press, pp. 25–32.

Aneo Keiji. (1988), "Butsu-Ya (Ki) no kōshō to tairitsu kōsō," in Nakao Takashi (ed.), *Ronshū Nihon bukkyōshi*, vol. 6, pp. 189–203. Tokyo: Yūzankaku.

Anesaki, Masaharu. (1963), *History of Japanese Religion*. Rutland, Vermont, and Tokyo: Tuttle (first edition 1930).

Antoni, Klaus. (1995), "The 'separation of Gods and Buddhas' at Ōmiwa Jinja in Meiji Japan," *Japanese Journal of Religious Studies*, 22/1–2 (Spring), 139–59.

Aoki Michio et al. (eds) (1981), *Seikatsu, bunka, shisō* (Ikki 4). Tokyo: Tōkyō Daigaku Shuppankai.

Asao Naohiro. (1970–4), "'Shōgun kenryoku' no sōshutsu" part 1, *Rekishi hyōron* 241 (August 1970), 70–8; part 2, *Rekishi hyōron* 266 (August 1972), 46–59; part 3, *Rekishi hyōron* 293 (September 1974), 20–36.

Bacon, Charlotte. (1932), "The call of Christ to China's daughters," *India's Women & China's Daughters*, November, 217–20.

Barbier, Jean-Paul. (1994), "The responsible and the irresponsible: Observations on the destruction and preservation of Indonesian art," in Paul Michael Taylor (ed.), *Fragile Traditions: Indonesian Art in Jeopardy*. Honolulu: University of Hawai'i Press, pp. 59–70.

Barclay, J. Gurney. (1937), "The missionary outlook in China," *India's Women & China's Daughters*, November, 175–7.

Barthes, Roland. (1972), *Mythologies*. New York: Farrar, Straus and Giroux.

—. (1973), *Elements of Semiology*. New York: Hill and Wang.

—. (1982), *A Barthes Reader*. Edited by Susan Sontag. New York: Hill and Wang.

Baudrillard, Jean. (1993), *Symbolic Exchange and Death*. London/Thousand Oaks, CA/New Delhi: Sage Publications.

—. (1996), *The System of Objects*. London and New York: Versus.

Beijing No. 26 Middle School Red Guards. (1996), "One hundred items for destroying the old and establishing the new," in Michael Schoenhals (ed.), *China's Cultural Revolution, 1966–1969: Not a Dinner Party*. Armonk, NY: M. E. Sharpe, pp. 212–22.

Belcher, Michael. (1991), *Exhibitions in Museums*. Leicester: Leicester University Press.

Belpoliti, Marco. (2002), *Crolli*. Turin: Einaudi.

Benjamin, Walter. (1968), *Illuminations*. New York: New York, Harcourt, Brace and World.

Benn, James A. (2007), *Burning for the Buddha: Self-immolation in Chinese Buddhism*. Honolulu: University of Hawai'i Press.

Besançon, Alain. (2000), *The Forbidden Image: An Intellectual History of Iconoclasm*. Chicago: University of Chicago Press.

Bettersby, C. Maud. (1897), "China's darkness and dawn," *India's Women and China's Daughters*, Part I, November, 249–51; Part II, December, 270–3.

Bird, J. G. (1922), "Hopes and fears in Chuki," *Church Missionary Outlook*, August, 170–1.

Bland, F. E. (1901), "The sub-prior of Kushan monastery," *Church Missionary Gleaner*, June, 87.

Blenkinsop, A. P. (1935), "Progress," *India's Women & China's Daughters*, July, 127–9.

Blissett, Luther. (2005), *Q*. New York: Mariner Press.

Blum, Mark L. (2006), "Rennyo Shōnin, manipulator of icons," in Blum and Yasutomi (eds), *Rennyo and the Roots of Modern Japanese Buddhism*. New York: Oxford University Press, pp. 109–36.

Blum, Mark L. and Yasutomi Shin'ya (eds) (2006), *Rennyo and the Roots of Modern Japanese Buddhism*. New York: Oxford University Press.

Boardman, Eugene Powers. (1952), *Christian Influence upon the Ideology of the T'ai-P'ing Rebellion, 1851–1864*. Madison: University of Wisconsin.

Boaz, Maud Elizabeth. (1912), "Days of reform in China," *India's Women & China's Daughters*, November, 215–17.

—. (1926), *And the Villages Thereof*. London: Morgan & Scott.

Bogel, Cynthea. (2002), "Canonizing Kannon: The ninth-century Esoteric Buddhist altar at Kanshinji," *Art Bulletin* (March), 30–64.

Boldrick, Stacy and Richard Clay (eds) (2007), *Iconoclasm: Contested Objects, Contested Terms*. Aldershot, Hampshire, England: Ashgate.

Boon, James A. (1991), "Why museums make me sad," in Ivan Karp and Steven D. Lavine (eds), *Exhibiting Cultures: The Poetics and Politics of Museum Display*. Washington, DC: Smithsonian Institution Press, pp. 255–77.

Boreham, F. (1936), "The missionary outlook in China," *India's Women & China's Daughters*, November, 191–6.

Boucher, Daniel (trans.) (1995), "Sūtra on the merit of bathing the Buddha," in Donald S. Lopez, Jr. (ed.), *Buddhism in Practice*. Princeton: Princeton University Press, pp. 58–68.

Bourdieu, Pierre. (1990), *The Logic of Practice*. Stanford: Stanford University Press.

—. (1991), *Language and Symbolic Power.* Cambridge: Harvard University Press.
Bourdieu, Pierre and Loic J. D. Wacquant. (1992), *An Invitation to Reflextive Sociology.* Chicago: University of Chicago Press.
Bracken, Catherine Philippa. (1975), *Antiquities Acquired: The Spoliation of Greece.* Newton Abbot: David & Charles.
Brawne, Michael. (1982), *The Museum Interior: Temporary and Permanent Display Techniques.* New York: Architectural Book Publishing Co.
Brown, Delmer M. and Ichirō Ishida (trans.) (1979), *The Future and the Past: A Translation and Study of the Gukanshō, an Interpretative History of Japan Written in 1219.* Berkeley: University of California Press.
Brown, Robert L. (1998), "The miraculous Buddha image: Portrait, god, or object?" in Richard H. Davis (ed.), *Images, Miracles, and Authority in Asian Religious Traditions.* Boulder: Westview Press, pp. 37–54.
"Buddhist idolatry in China." (1847), *Church Missionary Gleaner.* August, 92–3.
Burkert, Walter. (1983), *Homo Necans: The Anthropology of Ancient Greek Sacrificial Ritual and Myth.* Berkeley and Los Angeles: University of California Press.
Burroughs, F. R. (1900), "How God is working in Fuh-kien," *India's Women and China's Daughters*, August, 188.
Bush, Charles P. (1865), *Five Years in China: or, The Factory Boy made a Missionary. The Life and Observations of Rev. William Aitchison, Late Missionary to China.* Philadelphia: Presbyterian Board of Publication.
Bush, Richard C., Jr. (1970), *Religion in Communist China.* Nashville and New York: Abingdon Press.
Bussho kankōkai (ed.) (1978), *Dai Nihon bukkyō zensho.* 150 vols. Tokyo: Meicho fukyōkai (original edition 1912–22).
Callum, D. A. (1907), "In due season ye shall reap," *Church Missionary Gleaner*, December, 188.
Calvin, Jean. (1960), *Institutes of the Christian Religion.* Louiseville, KY: Westminster Press.
Candea, Virgil. (1987), "Iconoclasm," in Mircea Eliade (ed.), *Encyclopedia of Religion.* London: Collier Macmillan, Vol. 7, pp. 1–2.
Carter, Martha L. (1990), *The Mystery of the Udayana Buddha.* Supplemento n. 64 agli *Annali* vol. 50, fasc. 3. Naples: Istituto Universitario Orientale.
"Cases of conversion to Christianity at the Cities of Lo-nguong in the Provinces of Fuh-chau, China," (1867), *Church Missionary Gleaner*, July, 79–81.
Chapman, M. E. (1927), "Difficult days in Hengchow," *India's Women & China's Daughters*, June, 118–19.
Chiba Jōryū. (1990), "Rennyo no ikonokurasumu," in Chiba Jōryū hakase koki kinenkai (eds), *Nihon no shakai to bukkyō.* Tokyo: Nagata Bunshōdō, 1990, pp. 1–54.
"China – Distress of nations with perplexity," (1862), *Church Missionary Gleaner*, July, 77–9.
"Chinese justice," (1853), *Church Missionary Gleaner*, March, 31–3.
"Chinese missionary work," (1851), *Church Missionary Gleaner*, January, 111–13.

Chu Youyu. (1998), "Wozai 1966 nian," in Chu Youyu (ed.), *1966 women nayidai de huiyi*. Beijing: Zhongguo wenlian chubanshe, pp. 17–41.

Clarke, Prescott and Gregory, J. S. (1982), *Western Reports on the Taping: A Selection of Documents*. Honolulu: University of Hawai'i.

Codrington (Miss). (1915), "Chinese idols of gilt and mud," *Homes of the East*, n. 44, January, 7.

Cohen, Paul A. (1963), *China and Christianity: The Missionary Movement and the Growth of Chinese Antiforeignism, 1860–1870*. Cambridge: Harvard University Press.

Conger, Sarah Pike. (1909), *Letters from China: With Particular Reference to the Empress Dowager and the Women of China*. Chicago: A. C. McClurg.

Cook, A. E. (1922), "A Chinese master's influence," *Church Missionary Outlook* (August 1922), 171.

Coomaraswamy, Ananda. (1972), *The Origin of the Buddha Image*. New Delhi: Munshiram Manoharlal (original edition 1927).

Cooper, F. M. (1924), "Old saints of Loyuan," *India's Women and China's Daughters*, August–September, 142–4.

"Co-operation in war service," (1938), *Church Missionary Outlook*, November, 256.

Corless, Roger. (1996), "Idolatry and inherent existence: The golden calf and the wooden Buddha," in David Loy (ed.), *Healing Deconstruction: Postmodern Thought in Buddhism and Christianity*. Atlanta: Scholars Press, pp. 11–23.

Couche, Edith. (1915), "The Kwangsi-Hunan mission," *India's Women & China's Daughters*, October, 184–7.

—. (1916), "The swing of the pendulum," *Church Missionary Gleaner*, July, 93.

Culler, Jonathan. (1988), *Framing the Sign. Criticism and Its Institutions*. Oxford: Blackwell.

Daiki Naohiko. (1995), "Butsuzō no shōshitsu," *Rekishigaku kenkyū*, 9, 1995, 1–17.

Daitō Satoshi. (1994), *Otera no kane wa naranakatta: Bukkyō no sensō sekinin wo tou*. Tokyo: Kyōiku shiryō shuppankai.

Darley, Mary E. (1908), "Two seekers," *India's Women and China's Daughters*, February, 28–30.

—. (1914), "While it is day," *India's Women & China's Daughters*, January, 11–13.

—. (1930), "To-day in Kienning," *India's Women & China's Daughters*, March, 45–8.

Davis, Natalie Zemon. (1975), *Society and Culture in Early Modern France*. Stanford, CA: Stanford University Press.

Davis, Richard H. (1997), *Lives of Indian Images*. Princeton: Princeton University Press.

deBary, Wm. "Theodore and JaHyun Kim Haboush," (eds) (1985), *The Rise of Neo-Confucianism in Korea*. New York: Columbia University Press.

de Certeau, Michel. (1984), *The Practice of Everyday Life*. Berkeley: University of California Press.

Demiéville, Paul. (1974), "L'iconoclasme anti-bouddhique en Chine," in *Mélanges d'histoire des religions offerts à Henri-Charles Puech*. Paris: Presses Universitaires de France, pp. 17–25.

Deuchler, Martina. (1992), *The Confucian Transformation of Korea: A Study of Society and Ideology*. Cambridge, MA: Council on East Asian Studies, Harvard University and Harvard University Press.

Duara, Prasenjit. (1995), *Rescuing History from the Nation: Questioning Narratives of Modern China*. Chicago: University of Chicago Press.

Duffy, Aemon. (1992), *The Stripping of the Altars: Traditional Religion in England, 1400–1580*. New Haven: Yale University Press.

Duncan, Carol. (1995), *Civilizing Rituals: Inside Public Art Museums*. New York: Routledge.

Durand, Gilbert. (1987), "The imaginal," in Mircea Eliade (ed.), *Encyclopedia of Religion*. London: Collier Macmillan, Vol. 7, pp. 109–14.

Eco, Umberto. (1975), *A Theory of Semiotics*. Bloomington: Indiana University Press.

—. (1980), *La struttura assente: La ricerca semiotica e il metodo strutturale*. Milan: Bompiani (original edition 1968).

Eire, Carlos M. N. (1996), "Iconoclasm," in Hans J. Hillerbrand (ed.), *The Oxford Encyclopedia of the Reformation*, Vol. 2, pp. 302–6.

Ekeland, T., Arderson, Albert, and Christensen, Olive T. (eds) (1934), *White Unto Harvest in China: A Survey of the Lutheran United Mission, the China Mission of the N.L.C.A. 1890–1934*. Minneapolis: The Board of Foreign Missions.

Eliade, Mircea. (1959), *The Sacred and the Profane: The Nature of Religion*. New York: Harcourt Brace.

Elison, George. (1973), *Deus Destroyed: The Image of Christianity in Early Modern Japan*. Cambridge, MA and London: Harvard University Press.

Elwin, Arthur. (1903), "Idols' souls," *Church Missionary Gleaner*, April, 61.

"Encouragement and opposition," (1925), *Church Missionary Outlook*, 81–2.

Eubanks, Charlotte. (2011), *Miracles of Book and Body*. Berkeley: University of California Press.

Faure, Bernard. (1991), *The Rhetoric of Immediacy*. Princeton: Princeton University Press.

—. (1996), *Visions of Power: Imagining Medieval Japanese Buddhism*. Princeton, NJ: Princeton University Press.

—. (1998), "The Buddhist icon and the modern gaze," *Critical Inquiry*, 24/3, 768–813.

Fenollosa, Ernest F. (1911), *Epochs of Chinese and Japanese Art*. Vol. 1. New York: Frederick A. Stokes Co.

Feuchtwang, Stephan. (1989), "The problem of 'superstition' in the People's Republic of China," in Gustavo Benavides and M. W. Daly (eds), *Religion and Political Power*. Albany: State University of New York Press, pp. 43–68.

Flood, Finbarr Barry. (2002), "Between cult and culture: Bamiyan, Islamic iconoclasm, and the museum," *The Art Bulletin*, 84/4, 641–59.

Florovsky, Georges. (1974), *Christianity and Culture*. Belmont, MA: Nordland.

"Foochow—Extracts from recent letters," (1895), *India's Women & China's Daughters*, October, 474.

Fortesque, Adrian. (1913–1922), "Iconoclasm," in James Hastings (ed.), *Encyclopædia of Religion and Ethics*. Edinburgh: T & T Clark, Vol. 7, pp. 78–81.

Fortune, Robert. (1853), *Two Visits to the Tea Countries of China*. 2 vols. London: John Murray.

—. (1857), *A Residence among the Chinese: Inland, on the Coast, and at Sea*. London: John Murray.

Foster, Mrs. Arnold. (1899), *In the Valley of the Yangtse*. London: London Missionary Society.

Foucault, Michel. (1979), *Discipline and Punish: The Birth of the Prison*. Translated by Alan Sheridan. New York: Vintage Books.

Franck, Harry A. (1923), *Wandering in Northern China*. New York and London: Century Co.

Freedberg, David.(1985), *Iconoclasts and Their Motives*. Maarssen: Gary Schwartz.

—. (1989), *The Power of Images: Studies in the History and Theory of Response*. Chicago: University of Chicago Press.

Gamboni, Dario. (1997), *The Destruction of Art: Iconoclasm and Vandalism since the French Revolution*. New Haven: Yale University Press.

Geary, Patrick. (1994), *Living with the Dead in the Middle Ages*. Ithaca, NY: Cornell University Press.

Geertz, Clifford. (1973), *The Interpretation of Cultures*. New York: Basic Books.

Gell, Alfred. (1998), *Art and Agency: An Anthropological Theory*. Oxford: Clarendon Press.

Gernet, Jacques. (1985), *China and the Christian Impact: A Conflict of Cultures*. Translated by Janet Lloyd. Cambridge: Cambridge University Press; and Paris: Editions de la Maison des Sciences de l'Homme.

—. (1995), *Buddhism in Chinese Society: An Economic History from the Fifth to the Tenth Centuries*. New York: Columbia University Press.

Girard, René. (1977), *Violence and the Sacred*. Baltimore: Johns Hopkins University Press.

Gjertson, Donald E. (1989), *Miraculous Retribution: A Study and Translation of T'ang Lin's Ming-pao chi*. Berkeley Buddhist Studies Series. Berkeley: University of California Press.

Goffman, Erving. (1967), *Interactional Ritual: Essays on Face-to-face Behavior*. Garden City, New York: Anchor.

Goforth, Rosalind and Jonathan. (1931), *Miracles Lives of China*. New York and London: Harper & Brothers.

Gonoi Takashi. (1990), *Nihon kirisutokyō shi*. Tokyo: Yoshikawa Kōbunkan.

Goodchild, T. C. (1904), "An aged Buddhist priest," *Church Missionary Gleaner*, January, 11–12.

Goodwin, Janet R. (1994), *Alms and Vagabonds: Buddhist Temples and Popular Patronage in Medieval Japan*. Honolulu: University of Hawai'i Press.

Goossaert, Vincent. (2006), "1898: The beginning of the end for Chinese religion?" *Journal of Asian Studies*, 62/2, 307–35.

Gotō Tanji and Kamada Kisaburō (eds.) (1960–2), *Taiheiki*. 3 Vols (Nihon koten bungaku taikei 34–6). Tokyo: Iwanami shoten.

Goulde, John Isaac. (1985), "Anti-Buddhist polemic in fourteenth and fifteenth century Korea: The emergence of Confucian exclusivism." PhD dissertation, Harvard University.

Graham, Masako Nakagawa. (1998), *The Yang Kuei-fei Legend in Japanese Literature*. Lewiston: Mellen.

Grapard, Allan G. (1982), "Flying mountains and walkers of emptiness," *History of Religions*, 21/3 (February 1982), 195–221.

—. (1984), "Japan's ignored cultural revolution: The separation of Shinto and Buddhist divinities (*shinbutsu bunri*) in Meiji and a case study, Tonomine," *History of Religions*, 23/3, 240–65.

—. (1992), *The Protocol of the Gods: A Study of the Kasuga Cult in Japanese History*. Berkeley: University of California Press.

—. (1994a), "Rites de voyage: Redressing Hachiman's appearance," in *Mélanges Offerts en l'Honneur de René Sieffert*. Special issue of *Cipango: Cahiers d'études japonaises*, June 1994. Paris: Institut National des Langues et Civilizations Orientales, Centre d'Etudes Japonaises, pp. 335–52.

—. (1994b), "Geosophia, geognosis, and geopiety: Orders of significance in Japanese representations of space," in Deirdre Boden and Roger Friedland (eds), *NowHere: Time, Space and Modernity*. Berkeley: University of California Press, 1994, pp. 372–401.

—. (1998), "Geotyping sacred space: The case of Mount Hiko in Japan," in Benjamin Z. Kedar and R. J. Zwi Werblowsky (eds), *Sacred Space: Shrine, City, Land*. New York: New York University Press, 1998, pp. 215–49.

—. (2003), "The source of oracular speech: absence? Presence? Or plain treachery?" in Teeuwen and Rambelli (eds), *Buddhas and Kami in Japan*. London: Routledge Curzon, pp. 77–94.

Griggs, K. M. (1923), "The story of Ching Kang," *India's Women & China's Daughters*, August, 117–19.

—. (1929), "Scraps of information from a mountain tour," *India's Women & China's Daughters*, August, 155–6.

Groner, Paul. (2002), *Ryōgen and Mount Hiei: Japanese Tendai in the Tenth Century*. Honolulu: University of Hawai'i Press.

Gutzlaff, Charles. (1832), *Journal of Three Voyages along the Coast of China, in 1831, 1832, & 1833, with Notices of Siam, Corea, and the Loo-Choo Islands*. London: Frederick Westley and A. H. Davis.

Hague, Eric. (1947), "A school for Christian commandos," *Church Missionary Outlook*, December, 3.

Hakeda, Yoshito S. (1972), *Kūkai: Major Works*. New York: Columbia University Press.

Hall, Donald E. (ed.) (1994), *Muscular Christianity: Embodying the Victorian Age*. Cambridge: Cambridge University Press.

Hamerton-Kelly, Robert G. (ed.) (1987), *Violent Origins: Walter Burkert, René Girard, and Jonathan Z. Smith on Ritual Killing and Cultural Formations*. Stanford: Stanford University Press.

Hanawa Hokiichi and Ōta Tōshirō (eds) (1959–60), *Gunsho ruijū*. 29 vols. Tokyo: Zoku Gunsho ruijū kanseikai.

Hansen, Valerie. (1990), *Changing Gods in Medieval China, 1127–1276*. Princeton: Princeton University Press.

Hayami Tasuku. (1991), "Heian makki no jidai to shūkyō," *Chisan gakuhō*, 54 (March), 1–29.

Heine, Steven. (1994), "Sōtō Zen and the Inari cult," *Pacific World*, New Series, 10 (Fall), 75–101.

Henderson, Gregory and Hurvitz, Leon (1956), "The Buddha of Seiryoji: New finds and new theory," *Aribus Asiae*, 19/1, 5–55.

Hevia, James. (1995), *Cherishing Men From Afar: Qing Guest Ritual and the Macartney Embassy of 1793*. Durham: Duke University Press.

Hoare, J. C. (1894), "A temple at Tai-Chow," *Church Missionary Gleaner*, June, 91.

Hori, Ichirō. (1962), "Self-mummified Buddhas in Japan," *History of Religion*, 1/2, 222–42.

Horton, Sarah J. (2007), *Living Buddhist Statues in Early Medieval and Modern Japan*. New York: Palgrave Macmillan.

Houghton, Walter E. (1957), *The Victorian Frame of Mind, 1830–1870*. New Haven: Yale University Press.

Hua, Gu. (1985), *Pagoda Ridge and Other Stories*. Translated by Gladys Yang. Beijing: Panda Books.

Huc, Evariste Régis. (1855), *A Journey Through the Chinese Empire*. New York: Harper & Brothers, 2 vols.

Hurvitz, Leon (trans.) (1976), *Scripture of the Lotus Blossom of the Fine Dharma*. New York: Columbia University Press.

Høbjerg, Christian Kordt. (2002), "Inner iconoclasm: Forms of reflexivity in Loma rituals of sacrifice," *Social Anthropology*, 10/1, 57–75.

"In the track of the Communists," (1934), *Church Missionary Outlook*, November, 255.

Inoue, Takami. (2003), "The interaction between Buddhist and Shinto traditions at Suwa Shrine," in Teeuwen and Rambelli (eds), *Buddhas and Kami in Japan*. London: Routledge Curzon, pp. 287–312.

Ishizuka Masahide. (1993), *Fetishizumu no shinkōken: shinbutsu gyakutai no fōkuroā*. Tokyo: Sekai shoin.

Itoi, Kay and B. J. Lee. (2005), "A tussle over treasures," *Newsweek International*, February 21, 2005, 55–6.

Ives, Christopher. (1999), "The mobilization of doctrine: Buddhist contributions to imperial ideology in modern Japan," *Japanese Journal of Religious Studies*, 26/1–2 (Spring), 83–106.

—. (2001), "Protect the Dharma, protect the country: Buddhist war responsibility and social ethics," *The Eastern Buddhist*, 33/2, 15–34.

—. (2009), *Imperial-Way Zen: Ichikawa Hakugen's Critique and Lingering Questions for Buddhist Ethics*. Honolulu: University of Hawai'i Press.

Jackson, O. M. (1906), "Schools and churches in the western China mission," *Church Missionary Gleaner*, July, 108–10.
Jackson, S. M. (1912), "Awakening at Lo-Nguong," *India's Women & China's Daughters*, August, 157–8.
Jensen, Lionel M. (1997), *Manufacturing Confucianism: Chinese Traditions & Universal Civilization*. Durham: Duke University Press.
Jiang, Jili. (1997), *Red Scarf Girl: A Memoir of the Cultural Revolution*. New York: HarperCollins.
Jing, Jun. (1996), *The Temple of Memories: History, Power, and Morality in a Chinese Village*. Stanford, CA: Stanford University Press.
John Paul II. (1994), *Crossing the Threshold of Hope*. New York: Alfred A Knopf.
Kanda Chisato. (1991), *Ikkō ikki to Shinshū shinkō*. Tokyo: Yoshikawa Kōbunkan.
Kashiwahara Yūsen. (1990), *Nihon bukkyōshi: kindai*. Tokyo: Yoshikawa Kōbunkan.
Katō Shūichi (ed.) (1987–8). *Nihon sono kokoro to katachi*. 10 vols. Tokyo: Heibonsha.
Katsumata Shizuo. (1982), *Ikki*. Tokyo: Iwanami shoten.
Keefe, Patrick Radden. (2007), "The idol thief," *The New Yorker*, May 7, 58–67.
Keene, Donald (ed.) (1955), *Anthology of Japanese Literature from the Earliest Era to the Mid Nineteenth-Century*. New York: Grove Press (reprint 1981).
Keirstead, Thomas. (1992), *The Geography of Power in Medieval Japan*. Princeton, NJ: Princeton University Press.
Ketelaar, James E. 1990. *Of Heretics and Martyrs in Meiji Japan*. Princeton: Princeton University Press.
Kibbey, Ann. (1986), *The Interpretation of Material Shapes in Puritanism: A Study of Rhetoric, Prejudice, and Violence*. Cambridge: Cambridge University Press.
Kieschnick, John. (2003), *The Impact of Buddhism on Chinese Material Culture*. Princeton: Princeton University Press.
Kitagawa, Hiroshi and Bruce T. Tsuchida (trans.) (1975), *The Tale of the Heike*. 2 vols. Tokyo: University of Tokyo Press.
Kitagawa, Joseph. (1969), *Religion in Japanese History*. New York: Columbia University Press.
Kornicki, Peter F. (1994), "Public display and changing values: Early Meiji exhibitions and their precursors," *Monumenta Nipponica*, 49/2 (Summer), 167–96.
Kristeva, Julia. (1982), *Powers of Horror: An Essay on Abjection*. New York: Columbia University Press.
Kuroda Toshio. (1980), *Jisha seiryoku: mō hitotsu no chūsei shakai*. Tokyo: Iwanami shoten.
—. (1996), "The imperial law and the Buddhist law," *Japanese Journal of Religious Studies*, 23/3–4 (Fall), 271–85.
LaFleur, William R. (1990), "A turning in Taishō: Asia and Europe in the early writings of Watsuji Tetsurō," in J. Thomas Rimer (ed.), *Culture and Identity: Japanese Intellectuals during the Interwar Years*. Princeton: Princeton University Press, pp. 234–56.

Latour, Bruno. (2002), "What is iconoclash? Or is there a world beyond the image wars?" in Bruno Latour and Peter Weibel (eds), *Iconoclash: Beyond the Image Wars in Science, Religion, and Art*. Karlsruhe, Germany: ZKM, and Cambridge, MA: MIT Press, 2002, pp. 15–40.

Law, S. E. (1931), "The Church of China at war," *India's Women & China's Daughters*, June, 113–15.

Le Compte, Louis. (1787), *A voyage to, and description of CHINA, by Lewis Le Compte, with an Account of the Manners and Customs of its Inhabitants. Included in A Collection of Voyages and Travels*. Philadelphia: Spotswood.

Lee, Jonas. (1928), "Pakhoi in the absence of the missionaries," *Church Missionary Outlook*, January, 6–8.

Legge, James. (1965), *A Record of Buddhistic Kingdoms*. New York: Dover.

Lewis, W. J., Barber, William Theodore Aquila, and Hykes, John R. (eds) (1890), *Records of the General Conference of the Protestant Missionaries of China, Held at Shanghai, May 7–20, 1890*. Shanghai: American Presbyterian Mission Press.

Li, Rongxi (trans.) (1995), *A Biography of the Tripitaka Master of the Great Ci'en Monastery of the Great Tang Dynasty*. Berkeley: Numata Center for Buddhist Translation and Research.

— (trans.) (1996), *The Great Tang Dynasty Record of the Western Regions*. Berkeley: Numata Center for Buddhist Translation and Research.

Li, T. C. (Li Ganchen) and Lo, R. Y. (1924), *Pochu mixin quanshu*. Shanghai: The Mission Book Co.

Lincoln, Bruce. (1989), *Discourse and the Construction of Society: Comparative Studies of Myth, Ritual, and Classification*. Oxford: Oxford University Press.

Liscutin, Nicola. (2000), "Mapping the sacred body: Shinto versus popular beliefs at Mt. Iwaki in Tsugaru," in John Breen and Mark Teeuwen (eds), *Shinto in History*. Richmond, Surrey: Curzon Press, pp. 186–204.

Little, Lester K. (1993), *Benedictine Maledictions: Liturgical Cursing in Romanesque France*. Ithaca, NY and London: Cornell University Press.

Lloyd, Llewelyn (1900), "Five relics, and their story," *Church Missionary Gleaner*, August, 119.

Lopez, Donald S., Jr. (1998), *Prisoners of Shangri-La: Tibetan Buddhism and the West*. Chicago: University of Chicago Press.

Lotman, Yuri. (1990), *Universe of the Mind*. Bloomington: Indiana University Press.

Lutz, Jessie G and Rolland Ray Lutz. (1998), *Hakka Chinese Confront Protestant Christianity, 1850–1900: With the Autobiographies of Eight Hakka Christians, and Commentary*. Armonk, NY: M. E. Sharpe.

MacCallum, Donald. (1994), *Zenkōji and Its Icon*. Princeton: Princeton University Press.

Mackay, George Leslie. (1896), *From Far Formosa: The Island, Its People and Missions*. Edinburgh and London: Oliphant Anderson and Ferrier.

Main, D. Duncan. (1887), "Hospital work at Hang-Chow," *Church Missionary Gleaner*, January, 3.

Mangrum, Bryan D. and Giuseppe Scavizzi (trans. and intro.) (1991), *A Reformation Debate: Karlstadt, Emser, and Eck on Sacred Images: Three Treatises in Translation*. Toronto: Victoria University.

Mao Zedong. (1995), "Overthrowing the clan authority of the ancestral temples and clan elders, the religious authority of town and village gods, and the masculine authority of husbands," in Deborah A. Sommer (ed.), *Chinese Religion: An Anthology of Sources*. Oxford: Oxford University Press, pp. 305–8.

Marra, Michele. (1991), *The Aesthetics of Discontent: Politics and Reclusion in Medieval Japanese Literature*. Honolulu: University of Hawai'i Press.

Matsuyama Iwao. (1993), "'Kokuhō' to iu monogatari," in Geijutsu shinchō henshūbu (ed.), *Kokuhō*. Tokyo: Shinchōsha, pp. 177–92.

Maurer, Herrymon. (1951), *Collision of East and West*. Chicago: Henry Regnery.

McCullough, Helen Craig (trans.) (1959), *The Taiheiki. A Chronicle of Medieval Japan*. Rutland, VT and Tokyo: Tuttle (reprint 1985).

McMullin, Neil. (1984), *Buddhism and the State in Sixteenth-Century Japan*. Princeton, NJ: Princeton University Press.

—. (1988), "On placating the gods and pacifying the populace: The case of the Gion *Goryō* cult," *History of Religions*, 27/3 (February), 270–93.

—. (1989), "The *Lotus Sutra* and politics in the mid-Heian period," in George J. Tanabe, Jr., and Willa Jane Tanabe (eds), *The Lotus Sutra in Japanese Culture*. Honolulu: University of Hawai'i Press, 1989, pp. 119–41.

McNair, Amy. (2007), *Donors of Longmen: Faith, Politics, and Patronage in Medieval Chinese Buddhist Sculpture*. Honolulu: University of Hawai'i Press.

Medhurst, Walter Henry. (1836), "Extract from the manuscript journal of the Reverend W. H. Medhurst in the Huron, during her voyage along the eastern coast of China, in the summer and autumn of 1835," *Chinese Recorder*, Vol. 4 (January), 406–11.

—. (1846), "Missionary excursion from Shanghai," *Church Missionary Gleaner*, August, 90–2.

Merleau-Ponty, Maurice. (1962), *Phenomenology of Perception*. Translated by Colin Smith. New York: Humanities Press.

Michalski, Sergiusz. (1993), *The Reformation and the Visual Arts: The Protestant Image Question in Western and Eastern Europe*. London and New York: Routledge.

Mishima, Yukio. (1959), *The Temple of the Golden Pavilion*. New York: A.A. Knopf.

"Missionary labours at Ningpo, China," (1859), *Church Missionary Gleaner*, March, 28–31.

Mitter, Partha. (1977), *Much Maligned Monsters: A History of European Reactions to Indian Art*. Chicago: University of Chicago Press.

Miyasaka Yūshō (ed.) (1989), *Kōgyō Daishi chosakushū*. 2 vols. Tokyo: Sankibō busshorin (revised ed.).

Morrell, Robert E. (trans.) (1985), *Sand and Pebbles (Shasekishū): The Tales of Mujū Ichien, A Voice for Pluralism in Kamakura Buddhism*. Albany, NY: State University of New York Press.

Mott, John R. (1915), *The Present World Situation*. New York: Student Volunteer Movement for Foreign Missions.

Moule, Arthur Evans (1891), "China: and her woes and hopes," *Church Missionary Gleaner*, November, 163–4.

Murayama Shūichi. (1990), *Henbō suru kami to hotoketachi*. Kyoto: Jinbun shoin.

—. (2003). *Kyōto daibutsu goten seisuiki*. Kyoto: Hōzōkan.

Myers, James T. (1991), *Enemies Without Guns: The Catholic Church in the People's Republic of China*. New York: Paragon House.

Nagel's Encyclopedia Guide China. (1968), Translated by Anne L. Destenay. Geneva: Nagel Publishers.

Nattier, Jan. (1991), *Once Upon a Future Time: Studies in a Buddhist Prophecy of Decline*. Berkeley: Asian Humanities Press.

—. (trans.) (1995), "A prophecy of the death of the Dharma," in Donald S. Lopez, Jr. (ed.), *Buddhism in Practice*. Princeton: Princeton University Press, pp. 249–56.

Nicoloff, Philip. (2008), *Sacred Kōyasan: A Pilgrimage to the Mountain Temple of Saint Kobo Daishi and the Great Sun Buddha*. Albany: State University of New York Press.

Nihon mīra kenkyū gurūpu (eds) (1993), *Nihon, Chūgoku mīra shinkō kenkyū*. 2 vols. Tokyo: Heibonsha.

Nishikawa Joken. (1985), *Chōninbukuro, Hyakushōbukuro, Nagasaki yowakusa*. Tokyo: Iwanami shoten (original edition 1942).

Norton, E. M. (1930), "'Rejoice, o young man, in thy youth,'" *Church Missionary Outlook*, April, 70–3.

Oishio Chihiro. (1988), "Shiki-Hō seikenki no bukkyō seisaku to jiin," in Nakao Takashi (ed.), *Ronshū Nihon bukkyōshi*, Vol. 6, pp. 159–88. Tokyo: Yūzankaku.

Okakura Tenshin.(1939), *Nihon bijutsushi*. In *Okakura Tenshin zenshū*, edited by Okakura Kazuo, Vol. 4. Tokyo: Rokugeisha.

Okazaki Jōji (ed.) (1982), *Butsugu daijiten*. Tokyo: Kamakura shinsho.

Oliphant, Nigel. (1901), *A Diary of the Siege of the Legations in Peking During the Summer of 1900*. London: Longmans, Green & Co.

O'Neill, Mark. (1995), "Exploring the meaning of life: The St Mungo Museum of religious life and Art," *Museum International*, 185, 47/1, 50–3.

—. (1996), "Making histories of religion," in Gaynor Kavanagh (ed.), *Making Histories in Museums*. London: Leicester University Press, pp. 186–96.

"Our twenty-sixth anniversary meeting," (1906), *India's Women and China's Daughters*, June, 84–90.

"Outrage on the mission at Fuh-Chow," (1878), *Church Missionary Gleaner*, December, 137–8.

Pantin, Mabel. (1925), "Idols; or, the way," *Homes of the East*, April 1925, 14–16.

"Passing events in China," (1854), *Church Missionary Gleaner*, January, 2–4.

Perkins, George W. (trans.) (1998), *The Clear Mirror: A Chronicle of the Japanese Court During the Kamakura Period (1185–1333)*. Stanford, CA: Stanford University Press.

Phillips, Alice M. (1900), "Buddhist chanters in Kien-Ning," *India's Women and China's Daughters*, January, 16–17.

Phillips, H. S. (1893), "China missionaries in peril," *Church Missionary Gleaner*, January 3.

Phillips, John. (1973), *The Reformation of Images: Destruction of Art in England, 1535–1660*. Berkeley: University of California Press.

Pittman, Don A. (2001), *Toward a Modern Chinese Buddhism: Taixu's Reforms*. Honolulu: University of Hawai'i Press.

Pomian, Krzysztof. (1987), *Collectionneurs, amateurs et curieux. Paris, Venise: XVIe–XVIIIe siècle*. Paris: Gallimard.

—. (1994), "The collection: between the visible and the invisible," in Susan M. Pearce (ed.), *Interpreting Objects and Collections*. London and New York: Routledge, pp. 160–74.

—. (1997), "Histoire culturelle, histoire des sémiophores," in Jean-Pierre Rioux and Jean-Francois Sirinelli (eds), *Pour une histoire culturelle*. Paris: Seuil, pp. 83–100.

Porter, Bill. (1993), *Road to Heaven: Encounters with Chinese Hermits*. San Francisco: Mercury House.

Powell, William. (1991), "More laughter at Tiger Creek: Impressions of Buddhism in modern China," *Journal of Chinese Religion*, 12, 77–88.

Preziosi, Donald. (1986), *Iconoclasts and their Motives*. Montclair, NJ: G. Schwartz.

—. (1989), *Rethinking Art History: Meditations on a Coy Science*. New Haven: Yale University Press.

Prip-Moller, J. (1991), *Chinese Buddhist Monasteries: Their Plan and Its Function as a Setting for Buddhist Monastic Life* (new edition). Hong Kong: Hong Kong University Press.

Rambelli, Fabio. (2001), "Seizō hakai, kigō hakai, seisei hakai: seinaru mono no hakai kōi ni kansuru ichikōsatsu," *Hikaku bunka ronsō*, 8, September, 15–43.

—. (2002), "Secret Buddhas (*Hibutsu*): The limits of Buddhist representation," *Monumenta Nipponica*, 57/3, Autumn, 271–307.

—. (2004), "'Just behave as you like': Radical Amida cults and popular religiosity in premodern Japan," in Richard K. Payne and Kenneth K. Tanaka (eds), *Approaching the Land of Bliss: Religious Praxis in the Cult of Amitābha*. Honolulu: University of Hawai'i Press, pp. 169–201.

—. (2007), *Buddhist Materiality: A Cultural History of Objects in Japanese Buddhism*. Stanford, CA: Stanford University Press.

—. (2010a), "Home Buddhas: Historical processes and modes of representation of the sacred in the Japanese Buddhist family altar (*butsudan*)," *Japanese Religion*, 35/1–2 (Spring and Fall), 63–86.

—. (2010b), "The dangerous *kami* called Buddha: Ancient conflicts between Buddhism and local cults and medieval attempts at resolution," *Pacific World*, Third series, 12, 147–69.

Rambelli, Fabio and Reinders, Eric. (2007), "What does iconoclasm create? What does preservation destroy? Some observations from case studies of China and Japan," in Stacy Boldrick and Richard Clay (eds), *Iconoclasm: Contested Objects, Contested Terms*. Aldershot, Hampshire, England: Ashgate, pp. 15–34.

Reader, Ian and George J. Tanabe, Jr. (1998), *Practically Religious: Worldly Benefits and the Common Religion of Japan*. Honolulu: University of Hawai'i Press.

Reinders, Eric. (1997), "The iconoclasm of obeisance: Protestant images of Chinese religion and the Catholic Church," *Numen*, XLIV/3, 296–322.

—. (2004a), *Borrowed Gods and Foreign Bodies: Christian Missionaries Imagine Chinese Religion*. Berkeley: University of California Press.

—. (2004b), "Lun fanfo pohaizhong de foxiang pohuai," in Sarah Fraser (ed.), *Siyuan caifu yu shesu gongliang*. Shanghai: Shanghai Fine Arts Press, pp. 57–63.

—. (2004c), "Monkey kings make havoc: Iconoclasm and murder in the Chinese Cultural Revolution," *Religion*, 34/3, 191–209.

—. (2005), "Recycling icons and bodies in Chinese anti-Buddhist persecutions," *Res: Anthropology and Aesthetics*, 48 (Autumn), 71–8.

—. (2010), "The Chinese macabre in missionary publications and horror fiction," in Anthony Clark (ed.), *Beating Devils and Burning Their Books: Views of China, Japan, and the West*. Ann Arbor: Association for Asian Studies and University of Michigan, pp. 15–41.

Reischauer, Edwin O. (trans.) (1955), *Ennin's Diary: The Record of a Pilgrimage to China in Search of the Law*. New York: Ronald Press.

"Resolutions of the C.M.S. General Committee, August 13th, 1895," *Church Missionary Gleaner*, September 1895, 142.

"Revolting scenes in the beggars' square, Canton—Cruelty of heathenism," (1845), *Church Missionary Gleaner*, December, 136–7.

Ricci, Matteo S. J. (1985), *The True Meaning of the Lord of Heaven (T'ien-chu Shih-i)*. Translated by Douglas Lancashire and Peter Hu Kuo-chen, S. J.; Chinese-English edition edited by Edward J. Malatesta, S. J. Variétés Sinologiques New Series 72. Taipei: Ricci Institute.

Rigg, John (1892), "The riots near Nien-king," *Church Missionary Gleaner*, September, 138.

Robson, James. (2003), "A Tang Dynasty Chan mummy [roushen] and a modern case of furta sacra? Investigating the contested bones of Shitou Xiqian," in Bernard Faure (ed.), *Chan Buddhism in Ritual Context*. London: Routledge, pp. 151–78.

Rodd, H. R. and L. J. Bryer. (1906), "After furlough: At Ciong-Bau once more," *India's Women and China's Daughters*, August, 124–5.

"Romanism at Chusan," (1852), *Church Missionary Gleaner*, January, 10–12.

Rossi-Landi, Ferruccio. (1968), *Il linguaggio come lavoro e come mercato*. Milano: Bompiani.

Runciman, Steven. (1977), *The Byzantine Theocracy*. Cambridge: Cambridge University Press.

Ruppert, Brian. (2000), *Jewel in the Ashes: Buddha Relics and Power in Early Medieval Japan*. Cambridge: Harvard University Press.

Saeki Etatsu. (2003), *Haibutsu kishaku hyakunen*. Miyazaki: Kōmyakusha.
Sasaki Hōsei. (1999), *Hiraizumi Chūsonji: Konjikidō to kyō no sekai*. Tokyo: Yoshikawa Kōbunkan.
"Scene of the Ku-Cheng massacre," (1895), *Church Missionary Gleaner*, November, 165.
"Schools in temples—Idols hidden by blackboards," (1907), *Church Missionary Gleaner*, December, 188–9.
Schopen, Gregory. (1997), *Bones, Stones, and Buddhist Monks*. Honolulu: University of Hawai'i Press.
—. (2005), *Figments and Fragments of Mahayana Buddhism in India*. Honolulu: University of Hawai'i Press.
Sebald, Wilfried Georg (2002), "A natural history of destruction," *New Yorker*, November 4, 66–77.
Seidel, Anna. (1981), "Note à propos du terme 'Trésor national' en Chine et au Japon," *Bulletin de l'Ecole Française d'Extrême-Orient*, 69, 229–61.
—. (1983), "Imperial treasures and Taoist sacraments: Taoist roots in the apocrypha," in Michel Strickmann (ed.), *Tantric and Taoist Studies in Honor of R.A. Stein*. Brussels: Institut Belge des Hautes Etudes Chinoises, Vol. 2, pp. 291–371.
Sharf, Robert H. (trans.) (1996a), "The scripture in forty-two sections," in Donald S. Lopez, Jr. (ed.), *Religions of China in Practice*. Princeton: Princeton University Press, pp. 360–71.
—. (trans.) (1996b), "The scripture on the production of Buddha images," in Donald S. Lopez, Jr. (ed.), *Religions of China in Practice*. Princeton: Princeton University Press, pp. 261–7.
—. (1999), "On the allure of Buddhist relics," *Representations*, 66, 75–99.
—. (2005), "Ritual," in Donald Lopez Jr. (ed.), *Critical Terms for the Study of Buddhism*. Chicago: University of Chicago Press.
Shaw, C. (1907), "Day-dawn in a Chinese village," *Church Missionary Gleaner*, March, 42–3.
Shigematsu Akihisa. (1973), *Chūsei Shinshū shisō no kenkyū*. Tokyo: Yoshikawa Kōbunkan.
Shimizu Tadashi. (1999), "Gaisetsu: Butsugu to girei," in Shimizu Tadashi (ed.), *Butsugu jiten*. Tokyo: Tōkyōdō shuppan, pp. 5–39.
— (ed.) (1999), *Butsugu jiten*. Tokyo: Tōkyōdō shuppan.
Sing, Tsae-seng. (1929), "A letter from Bishop Sing, Assistant Bishop of Chekiang," *Church Missionary Outlook*, May, 95.
"Sketches of idol-worship. IV. The strength of idolatry as an agency of priestcraft," (1874), *Church Missionary Gleaner*, October, 117–20.
Smith, Jonathan Z. (1978), *Map is Not Territory: Studies in the History of Religions*. Leiden: Brill.
—. (1982), *Imagining Religion: From Babylon to Jonestown*. Chicago: University of Chicago Press.
—. (1987), *To Take Place: Toward Theory in Ritual*. Chicago: University of Chicago Press.

Sommer, Deborah A. (1994), "Images into words: Ming Confucian iconoclasm," *National Palace Museum Bulletin*, 14, 1–24.
Soper, Alexander. (1959), *Literary Evidence for Early Buddhist Art in China*. Ascona, Switzerland: Artibus Asiae.
Spence, Jonathan D. (1996), *God's Chinese Son: The Taiping Heavenly Kingdom of Hong Xiuquan*. New York: W. W. Norton.
Stewart, E. L. (1932), "Christ on the Chinese highways.—[Part] I," *Church Missionary Outlook*, February, 28–31.
Stille, Alexander. (1998), "Faking it: How the Italians have taught the Chinese to preserve their ancient masterpieces instead of copying them," *New Yorker*, June 15, 36–42.
Stock, Eugene. (1890), *The Story of the Fuh-Kien Mission of the Church Missionary Society*. London: Seeley, Jackson & Halliday.
Stocking, George W., Jr. (1985), "Essays on museums and material culture," in George W. Stocking, Jr. (ed.), *Objects and Others: Essays on Museum and Material Culture*. History of Anthropology, Vol. 3. Madison: University of Wisconsin Press, pp. 3–14.
Stone, Jacqueline I. (1999), *Original Enlightenment and the Transformation of Medieval Japanese Buddhism*. Honolulu: University of Hawai'i Press.
Strickmann, Michel. (1996), *Mantras et mandarins: Le bouddisme tantrique en Chine*. Paris: Gallimard.
Swearer, Donald J. (2004), *Becoming the Buddha*. Princeton, NJ: Princeton University Press.
Takakusu Junjirō and Watanabe Kaigyoku (eds) (1924–32), *Taishō shinshū daizōkyō*. 85 vols. Tokyo: Issaikyō kankōkai.
— (eds) (1924–35), *Taishō shinshū daizōkyō zuzōbu*. Tokyo: Taishō issaikyō kankōkai and Daizō shuppan.
Takatsuki Tadahisa and Yoshino Taiichirō. (2004). "Kōrai butsuga dare no mono," *Asahi Shinbun*, November 25, 27.
Takei Akio. (1998), *Nihon kodai bukkyō no bunkashi*. Tokyo: Yoshikawa Kōbunkan.
Tamamuro Fumio. (1977), *Shinbutsu bunri*. Tokyo: Kyōiku shuppan.
Tamura Yasuho (1987), "Meiji Ishin haibutsu kishaku no chihōteki tenkai to sono tokushitsu ni tsuite," in Ikeda Eishun (ed.), *Ronshū Nihon bukkyōshi*, 8, pp. 69–87. Tokyo: Yūzankaku.
Tanaka, Stefan. (1994), "Imaging history: Inscribing belief in the nation," *Journal of Asian Studies*, 53/1, February, 24–44.
Taussig, Michael T. (1992), *The Nervous System*. New York: Routledge.
—. (1997), *The Magic of the State*. New York: Routledge.
—. (1999), *Defacement: Public Secrecy and the Labor of the Negative*. Stanford: Stanford University Press.
Teeuwen, Mark and Fabio Rambelli (eds) (2003), *Buddhas and Kami in Japan*. London: Routledge Curzon.
Thal, Sarah. (2002), "Redefining the gods: Politics and survival in the creation of modern kami," *Japanese Journal of Religious Studies*, 29/3–4 (Fall), 379–404.

—. (2005). *Rearranging the Landscape of the Gods: The Politics of a Pilgrimage Site in Japan, 1573–1912*. Chicago: University of Chicago Press.

"The day of opportunity and the C.M.S," (1908), *Church Missionary Gleaner*, September, 132–3.

"The Fuh-Kien mission," (1884), *Church Missionary Gleaner*, September, 104–6.

"The influence of students" *Church Missionary Outlook*, June 1925, 125–6.

"The movement in China," (1853), *Church Missionary Gleaner*, October, 110–11.

Thompson, E. (1900), "A local riot in mid China," *Church Missionary Gleaner*, October, 148–50.

Thompson, Michael. (1994), "The filth in the way," in Susan M. Pearce (ed.), *Interpreting Objects and Collections*. Abingdon: Routledge, pp. 269–78.

"Tidings from Ningpo," (1862), *Church Missionary Gleaner*, April, 43–4.

Tsang, Carol R. (2007), *War and Faith: Ikko Ikki in Late Muromachi Japan*. Cambridge, MA: Harvard University Asia Center and Harvard University Press.

Tsuji Zennosuke. (1955), *Nihon bukkyōshi*. 10 Vols. Tokyo: Iwanami (reprint 1983).

—. 1984. *Nihon bukkyō no kenkyū*. 6 vols. Tokyo: Iwanami shoten.

Tsunoda, Ryusaku, Wm. Theodore de Bary, and Donald Keene (eds) (1964), *Sources of Japanese Tradition*. 2 vols. New York: Columbia University Press.

Turner, Victor. (1995), *The Ritual Process: Structure and Anti-Structure*. New York: Aldine de Gruyter.

Twitchett, Denis Crispin (1955–6), "Monastic estates in T'ang China," *Asia Major*, 5/1, 123–46.

Tyler, Royall (trans.) (1977), *Selected Writings of Suzuki Shōsan*. Ithaca, NY: China-Japan Program, Cornell University.

Umeda Yoshihiko. (1971–2), *Nihon shūkyō seido shi*. 4 Vols (revised edition). Tokyo: Tōsen shuppan.

Victoria, Brian. (1997), *Zen at War*. New York and Tokyo: Weatherhill.

Walsh, Kevin. (1992), *The Representation of the Past: Museums and Heritage in the Post-modern World*. London: Routledge.

Walsh, Michael. (2010), *Sacred Economies: Buddhist Monasticism and Territoriality in Medieval China*. New York: Columbia University Press.

Wandel, Lee Palmer. (1995), *Voracious Idols and Violent Hands: Iconoclasm in Reformation Zurich, Strasbourg, and Basel*. Cambridge: Cambridge University Press.

Wang Yarong (ed.) (1986), *Daxingshansi*. Xi'an: San Qin chubanshe.

Wang, Youqin. (1999), "Selective memory and the Chinese Cultural Revolution," Unpublished paper given at the AAS, Boston, March 13, 1999.

Wee, C. J. W.-L. (1994), "Christian manliness and national identity: The problematic construction of a racially 'pure' nation," in Donald E. Hall (ed.), *Muscular Christianity: Embodying the Victorian Age*. Cambridge: Cambridge University Press, pp. 66–88.

Wehrle, Edmund S. (1966), *Britain, China, and the Antimissionary Riots 1891–1900*. Minneapolis: University of Minnesota Press.

Wen, Jieruo (1995), "Living hell (from her memoirs)," in Joseph S. M. Lau and Howard Goldblatt (eds), *The Columbia Anthology of Modern Chinese Literature*. New York: Columbia University Press, pp. 703–9.

Wessinger, Catherine. (2000), *How the Millennium Comes Violently: From Jonestown to Heaven's Gate*. New York and London: Seven Bridges Press.

Willett, Gear. (1931), "Present conditions in China," *India's Women & China's Daughters*, April, 72–4.

Williams, S. Wells. (1907), *The Middle Kingdom: A Survey of the Geography, Government, Literature, Social Life, Arts, and History of the Chinese Empire and its Inhabitants*. New York: Charles Scribner's Sons.

Woodbridge, Samuel Isett. (1919), *Fifty Years in China*. Richmond VA and Texarkana, AK-TX: Presbyterian Committee of Publication.

Xu, Yun. (1988), *Empty Cloud: The Autobiography of the Chinese Zen Master*. Translated by Charles Luk. Longmead: Element Books.

Yamaguchi, Masao. (1991), "The poetics of exhibition in Japanese culture," in Ivan Karp and Steven D. Lavine (eds), *Exhibiting Cultures: The Poetics and Politics of Museum Display*. Washington, DC: Smithsonian Institution Press, pp. 57–67.

Yamamoto Hiroko. (1989), "Shin no mihashira to chūseiteki sekai (10): Chūsei no sangū jinji (ge): Myōken no Hoshidō-zō wo megutte," *Shunjū*, n. 312.

—. (1998), *Ijin: Chūsei Nihon no hikyōteki sekai*. Tokyo: Heibonsha.

Yamamoto, Kōshō (trans.) (1968), *The Words of St. Rennyo: Complete Translations of the Rennyoshonin-Goichidaiki-Kikigaki and the Anjinketsujosho*. Ube: Karinbunko.

Yang, Hsüan-chih. (1984), *A Record of Buddhist Monasteries in Lo-yang*. Translated by Yi-t'ung Wang. Princeton: Princeton University Press.

Yasumaru Yoshio. (1979), *Kamigami no Meiji Ishin*. Tokyo: Iwanami shoten.

Yasutomi, Shin'ya. (2006), "The life of Rennyo: A struggle for the transmission of Dharma," in Blum and Yasutomi (eds), *Rennyo and the Roots of Modern Japanese Buddhism*. New York: Oxford University Press, pp. 17–37.

Yates, Matthew Tyson, Nelson, R., and Barrett, E. R. (eds) (1879), Records of the General Conference of the Protestant Missionaries of China, Held at Shanghai, May 10–24, 1877. Shanghai: Presbyterian Mission Press.

Yiengpruksawan, Mimi Hall. (1991) "In my image: The Ichiji Kinrin statue at Chūsonji," *Monumenta Nipponica*, 46/3, 329–47.

—. (1998), *Hiraizumi: Buddhist Art and Regional Politics in Twelfth-Century Japan*. Cambridge, MA: Harvard University Press.

Yingsan (pseud.) (1983), "One day in Dazhu," in Sherman Cochran, Andrew C. K. Hsieh and Janis Cochran (eds), *One Day in China: May 21, 1936*. New Haven: Yale University Press, pp. 155–9.

Yoshida Kyūichi. (1970), *Nihon no kindai shakai to bukkyō*. Tokyo: Hyōronsha.

Yuan, Gao. (1987), *Born Red: A Chronicle of the Cultural Revolution*. Stanford: Stanford University.

Zhai, Zhenhua. (1992), *Red Flower of China*. New York: Soho.

Zhulin. (1992), *Foju cidian*. Taipei: Changchunshu shufang.

Index